AF327796

Linear CMOS RF Power Amplifiers for Wireless Applications

ANALOG CIRCUITS AND SIGNAL PROCESSING SERIES

Consulting Editor: Mohammed Ismail. *Ohio State University*

For other titles published in this series, go to
www.springer.com/series/7381

Paulo Augusto Dal Fabbro · Maher Kayal

Linear CMOS RF Power Amplifiers for Wireless Applications

Efficiency Enhancement and Frequency-Tunable Capability

 Springer

Dr. Paulo Augusto Dal Fabbro
École Polytechnique Fédérale
de Lausanne (EPFL)
1015 Lausanne
Switzerland
pauloau@gmail.com

Prof. Maher Kayal
École Polytechnique Fédérale
de Lausanne (EPFL)
1015 Lausanne
Switzerland
maher.kayal@epfl.ch

Series Editors:
Mohammed Ismail
205 Dreese Laboratory
Department of Electrical Engineering
The Ohio State University
2015 Neil Avenue
Columbus, OH 43210, USA

Mohamad Sawan
Electrical Engineering Department
École Polytechnique de Montréal
Montréal, QC, Canada

ISBN 978-90-481-9360-8   e-ISBN 978-90-481-9361-5
DOI 10.1007/978-90-481-9361-5
Springer Dordrecht Heidelberg London New York

Library of Congress Control Number: 2010930617

Cover design: eStudio Calamar S.L.

Printed on acid-free paper

Springer is part of Springer Science+Business Media (www.springer.com)

Preface

Advances in electronics have pushed mankind to create devices, ranging from incredible gadgets to medical equipment to spacecraft instruments. More than that, modern society is getting used to—if not dependent on—the comfort, solutions, and astonishing amount of information brought by these devices. One field that has continuously benefitted from those advances is the radio frequency integrated circuit (RFIC) design, which in its turn has promoted countless benefits to the mankind as a payback. Wireless communications is one prominent example of what the advances in electronics have enabled and their consequences to our daily life. How could anyone back in the eighties think of the possibilities opened by the wireless local area networks (WLANs) that can be found today in a host of places, such as public libraries, coffee shops, trains, to name just a few? How can a youngster, who lives this true WLAN experience nowadays, imagine a world without it?

This book deals with the design of linear CMOS RF Power Amplifiers (PAs). The RF PA is a very important part of the RF transceiver, the device that enables wireless communications. Two important aspects that are key to keep the advances in RF PA design at an accelerate pace are treated: efficiency enhancement and frequency-tunable capability. For this purpose, the design of two different integrated circuits realized in a 0.11 µm technology is presented, each one addressing a different aspect.

With respect to efficiency enhancement, the design of a dynamic supply RF power amplifier is treated, making up the material of Chaps. 2 to 4. The frequency-tunable capability is addressed in Chaps. 5 to 7, presenting the design of a tunable impedance matching network for use in multiband RF power amplifiers. This novel network is based on coupled inductors. The reader can jump directly to Chap. 5 if the interest is for the frequency-tunable PA. When background information is important to the understanding of the material being presented, the reader will be guided to the page of concern. In Chap. 1, a detailed outline of the book can be found. Chapter 8 summarizes the results obtained and conclusions that could be drawn from the work underlying this book. Additional information, regarding the measurement setups and the implementation of the impedance matching networks used in the characterization of the power amplifiers, can be found in the two appendices that close the book.

The material presented in this book is the offspring of a collaborative research between the Electronics Laboratory of the *École polytechnique fédérale de Lausanne* (EPFL) in Lausanne, Switzerland, and the Fujitsu Laboratories Ltd. in Yokohama, Japan.

Acknowledgments

Thanks to Yuu Watanabe and Kazuhiko Kobayashi, from Fujitsu Labs, for the support during all the collaboration project. Thanks also to Cédric Meinen, from EPFL, for his invaluable contributions to the design of the dynamic supply system.

Lausanne, Switzerland   Paulo Augusto Dal Fabbro
                                                                           Maher Kayal

Contents

1 Introduction 1
  1.1 Context 1
  1.2 Objectives 2
    1.2.1 Efficiency Enhancement 2
    1.2.2 Frequency-Tunable Capability 3
  1.3 Book Outline 3
  References 4

2 Efficiency Enhancement 5
  2.1 Introduction 5
  2.2 Basic Principles 5
    2.2.1 Linearity 5
    2.2.2 Efficiency 6
  2.3 Overview of the Main Efficiency-Enhancement Techniques 7
    2.3.1 Envelope Elimination and Restoration 8
    2.3.2 Dynamic Supply RF Power Amplifier 10
  2.4 Conclusion 14
  References 14

3 Design of the Dynamic Supply CMOS RF Power Amplifier 17
  3.1 Introduction 17
  3.2 High-Efficiency, Fast Modulator 18
    3.2.1 High-Speed Comparator 20
    3.2.2 Synchronous Switch 21
    3.2.3 LC Filter 24
  3.3 RF Power Amplifier 25
    3.3.1 Design Procedure 26
    3.3.2 Stability 29
  3.4 Dynamic Supply RF Power Amplifier 32
  3.5 Conclusion 37
  References 37

4 Measurement Results for the Dynamic Supply CMOS RF Power Amplifier ... 39
  4.1 Introduction ... 39
  4.2 Main Evaluation Board ... 39
  4.3 Modulator Characterization ... 42
    4.3.1 Envelope Detection and Processing ... 42
  4.4 RF Power Amplifier Characterization ... 45
    4.4.1 S-Parameters Measurement ... 45
    4.4.2 Single-Tone Measurements ... 49
  4.5 Dynamic Supply RF PA Characterization ... 50
    4.5.1 Two-Tone Measurements ... 50
    4.5.2 OFDM Measurements ... 55
  4.6 Conclusion ... 58
  References ... 59

5 Frequency-Tunable Capability ... 61
  5.1 Introduction ... 61
    5.1.1 Scenario ... 62
  5.2 Definitions ... 62
    5.2.1 Tuning Range ... 62
  5.3 Overview of the Existing Frequency-Tunable Techniques ... 63
    5.3.1 Broadband Techniques ... 63
    5.3.2 Narrowband Techniques ... 65
  5.4 Conclusion ... 72
  References ... 73

6 Design of the Frequency-Tunable CMOS RF Power Amplifier ... 77
  6.1 Introduction ... 77
  6.2 Output Impedance Matching ... 77
    6.2.1 The π-Matching Network ... 78
    6.2.2 Tunable Output Impedance Matching ... 82
  6.3 Input Impedance Matching ... 83
  6.4 Frequency-Tunable RF Power Amplifier ... 83
    6.4.1 RF Power Amplifier ... 83
    6.4.2 Coupled Inductors ... 84
    6.4.3 Control Circuit ... 88
    6.4.4 Complete System ... 95
  6.5 Conclusion ... 99
  References ... 99

7 Measurement Results for the Frequency-Tunable CMOS RF Power Amplifier ... 101
  7.1 Introduction ... 101
  7.2 Stand-Alone RF Power Amplifier Characterization ... 102
  7.3 Coupled Inductors Characterization ... 104
  7.4 Frequency-Tunable RF PA Measurement ... 108

7.5 Hybrid Implementation ... 114
  7.5.1 Measurement Results for the Hybrid PA ... 115
7.6 Analysis and Discussion of the Results ... 118
7.7 Conclusion ... 120
  References ... 121

8 Conclusion ... 123
8.1 Highlights ... 123
8.2 Main Contributions ... 124
  8.2.1 Efficiency Enhancement ... 124
  8.2.2 Frequency-Tunable Capability ... 125
  8.2.3 Impedance Matching ... 125
8.3 Future Work ... 125
  8.3.1 Efficiency Enhancement ... 125
  8.3.2 Frequency-Tunable Capability ... 126
  References ... 126

9 Appendix A: Measurement Setups and Additional Screenshots ... 127
9.1 Setups Used for the S-Parameter Measurements ... 127
9.2 Setup Used for the 2-Tone Measurements ... 127
9.3 Setup Used for the OFDM Measurements ... 132
9.4 Additional Screenshots ... 134
  References ... 135

10 Appendix B: Procedure for Impedance Matching of Printed-Circuit RF Amplifiers ... 137
10.1 Introduction ... 137
10.2 Procedure for Impedance Matching ... 138
  10.2.1 PCB Access Lines Design and Fabrication ... 140
  10.2.2 SOLT Calibration and Measurement ... 141
  10.2.3 TRL Calibration and Measurement ... 141
  10.2.4 Access Line Modeling and De-Embedding ... 142
  10.2.5 Matching Network Implementation ... 142
  10.2.6 Summarized Guidelines ... 143
10.3 Application Example ... 143
  10.3.1 First Step ... 143
  10.3.2 Second Step ... 143
  10.3.3 Third Step ... 145
  10.3.4 Fourth Step ... 146
  10.3.5 Fifth Step ... 148
10.4 Analysis of the Results ... 152
  References ... 155

Index ... 157

Notation

Abbreviations and Acronyms

acc. to	according to
ACLR	Adjacent Channel Leakage Ratio [dBc]
ACPR	Adjacent Channel Power Ratio [dBc]
Bip.	Bipolar
bps	bits per second
BST	Barium-Strontium-Titanate (ferroelectric material)
CMFB	Common-Mode FeedBack
Const.	Constant Supply
Conv.	Conventional
CW	Constant Wave
DAC	Digital-to-Analog Converter
Discr.	Discrete
DPD	Digital PreDistortion
DPI	Driving-Point Impedance
DSP	Digital Signal Processor
Dyn.	Dynamic Supply
EDGE	Enhanced Data rates for GSM Evolution
EER	Envelope Elimination and Restoration
Eff.	Efficiency
FET	Field Effect Transistor
FM	FerroMagnetic
FR-4	Flame Retardant type 4 (base material for PCBs)
GIC	Generalized Impedance Converter
GSG	Ground-Signal-Ground
GSM	Global System for Mobile communications
HBT	Hetero-junction Bipolar Transistor
HF	High Frequency
HFET	Heterojunction FET
LF	Low Frequency
LNA	Low-Noise Amplifier

LO   Local Oscillator
MEMS   Micro-Electro-Mechanical System
MESFET   MEtal Semiconductor FET
MIM   Metal-Insulator-Metal (capacitor)
Mod.   Modulator
MOS   Metal-Oxide-Semiconductor
NADC   North American Digital Cellular
PAE   Power-Added Efficiency [%]
PAPR   Peak-to-Average Power Ratio
PCB   Printed-Circuit Board
PDA   Personal Digital Assistant
PGS   Patterned Ground Shield
PIN p^+ intrinsic n^+ (diode)
PWM   Pulse-Width Modulation
QAM   Quadrature Amplitude Modulation
RFC   RF Choke
rms   Root Mean Square
SFG   Signal-Flow Graph
SMA   SubMiniature version A connector
SMB   SubMiniature version B connector
SOLT   Short-Open-Load-Thru
SP   Scattering Parameters
SRF   Self-Resonant Frequency
TCM   Trellis Coded Modulation
TR   Tuning Range [%]
TRL   Thru-Reflect-Line
Tun.   Tunable
VCO   Voltage-Controlled Oscillator
VSWR   Voltage Standing Wave Ratio

Greek Symbols

α	real part of the ratio between I_{ctrl} and I_{RF}
β	imaginary part of the ratio between I_{ctrl} and I_{RF}
ε	dielectric constant (or permittivity) [F/m]
ε_{r}	relative dielectric constant (or relative permittivity)
η_{d}	drain efficiency [%]
Γ_{IN}	input reflection coefficient
Γ_{L}	load reflection coefficient
Γ_{ML}	load reflection coefficient for simultaneous conjugate match
Γ_{MS}	source reflection coefficient for simultaneous conjugate match
Γ_{opt}	optimum reflection coefficient
Γ_{OUT}	output reflection coefficient
Γ_{S}	source reflection coefficient
λ	wavelength [m]
μ	electron or hole effective mobility [m^2/(V $\times$ s)]
ω	angular frequency [rad/s]
ω_{n}	undamped natural frequency [rad/s]
ϕ	phase of the ratio between I_{ctrl} and I_{RF} [$^\circ$ or rad]
ϕ_{LC}	phase shift introduced by the LC filter [$^\circ$ or rad]
ϕ_{td}	phase shift corresponding to t_d [$^\circ$ or rad]
ϕ_{total}	total modulator phase shift ($\phi_{\text{LC}} + \phi_{\text{td}}$) [$^\circ$ or rad]
ξ	damping ratio

Roman Symbols

b	ratio between the electron and hole effective mobilities
C_{ox}	MOS transistor's gate oxide capacitance per unit area [F/m^2]
d_{in}	inner diameter of an inductor or transformer [m]
d_{out}	outer diameter of an inductor or transformer [m]
f_c	center frequency [Hz]
f_s	modulator's switching frequency [Hz]
G	power gain [dB or W/W]
g_m	transconductance of a MOS transistor [A/V]
I_{ctrl}	current that controls the tunable inductor [A]
I_d	drain current of a MOS transistor [A]
I_{RF}	current flowing through the tunable inductor [A]
K	Rollett stability factor
k	coupling factor between two magnetically coupled inductors
L	self inductance or MOS transistor channel length [H or m]
M	mutual inductance between two coupled inductors [H]
m	transformation factor of a matching network
n	number of turns of an inductor winding
P_{DC}	DC power consumption [W]
P_{in}	input power [dBm or W]
P_{out}	output power [dBm or W]
Q	quality factor
Q_0	loaded quality factor of a matching network
Q_u	unloaded quality factor of an inductor
r	magnitude of the ratio between I_{ctrl} and I_{RF}
R_{Ls}	parasitic series resistance of an inductor [Ω]
r_{on}	on resistance of a MOS transistor [Ω]
R_{opt}	optimum resistance presented at the PA output [Ω]
s	track spacing for a planar inductor or transformer [m]
T	dielectric thickness [m]
t_d	modulator delay [s]
t_{ed}	electrical delay of a transmission line [s]
V_{knee}	turn-on drain voltage of a MOS transistor [V]
V_T	threshold voltage of a MOS transistor [V]
W	MOS transistor's channel width [m]
w	track width for a planar inductor or transformer [m]
Z_0	characteristic impedance of a system or transmission line [Ω]
Z_{opt}	optimum impedance [Ω]
Z_{PA}	impedance seen by the supply when looking to the PA [Ω]

Chapter 1
Introduction

Abstract This introductory chapter presents the motivation for the subjects covered by this book. They are the efficiency enhancement and the frequency tunable capability applied to linear CMOS RF power amplifiers for wireless applications. The objectives, the specifications, and the outline of the book are described.

1.1 Context

Modern communication systems call for high data rates. Sophisticated modulations schemes have been developed to overcome problems such as multipath that prevented data rates from being increased. One such scheme is the Orthogonal Frequency Division Multiplexing (OFDM) adopted by the IEEE 802.11a/g standards that regulate Wireless Local-Area Network (WLAN) applications at 2.4 and 5.2 GHz [1, 2]. This book addresses two aspects of Radio-Frequency (RF) Power Amplifiers (PAs) for application in WLANs.

The first aspect is efficiency enhancement. In variable envelope modulation schemes, like OFDM, the amplitude of the modulated signal contains essential information about the signal being transmitted. Differing from constant envelope modulation, for which the amplitude does not bear any signal content, variable envelope systems demand linear transmitters. Linearity comes at the price of reduced efficiency, which decreases the autonomy of battery-powered devices. This trade-off between efficiency and linearity must be circumvented, which translates into improving the performance of the PA at the front-end of the transmitter. The PA is always considered one of the key performance bottlenecks in RF systems because it is responsible for the major part of their DC power consumption. Applying an efficiency-enhancement technique to a power amplifier is a possible solution to overcome the linearity–efficiency trade-off.

The second issue to be dealt with is the possibility of employing a PA in mobile devices operating according to different standards so that the device in question can be used in different environments and situations. A multi-standard PA like this would require some flexibility relative to its frequency of operation. If this power

P.A. Dal Fabbro, M. Kayal, *Linear CMOS RF Power Amplifiers for Wireless Applications,*   1
Analog Circuits and Signal Processing,
DOI 10.1007/978-90-481-9361-5_1, © Springer Science+Business Media B.V. 2010

amplifier must comply with both IEEE 802.11a and 802.11g standards, for instance, it should be able to operate in both the 2.4 and 5.2 GHz bands. In this book, we call this feature the frequency-tunable capability.

Both issues make up part of the scenario of present and future local and personal area networks. Increased comfort and more liberty to the user can be achieved when these issues are addressed.

> **Units**
>
> The units of the *Système International* (SI) are used throughout this document. The power gain is always expressed in dB, whereas the primary unit for power is dBm and the secondary is W—always given in parentheses after dBm.

1.2 Objectives

1.2.1 Efficiency Enhancement

One of the objectives is to develop a Complementary Metal-Oxide-Semiconductor (CMOS) RF PA in which the linearity–efficiency trade-off is mitigated, so that it can be used in a system employing a variable envelope modulation scheme. To achieve this, the dynamic supply technique is used. It combines a high-efficiency, fast modulator operating in switched-mode and a linear PA. The effectiveness of this principle had already been demonstrated in a previous work [4] by Schlumpf. In his thesis, Schlumpf used a discrete power amplifier based on silicon bipolar transistors for Code-Division Multiple Access (CDMA) applications at 1.9 GHz. This book presents a system designed in a 0.11 μm CMOS technology, in which both the modulator and the PA are integrated on the same chip.

The interest of an implementation in CMOS of an existing solution in bipolar technology resides mainly on the pursuit of the so-called single chip radio. Such an integrated circuit (IC) would allow increased functionality to be added to mobile devices since more room is freed by the integration of all the parts of a transceiver on the same die. The integration of a whole transceiver implies its realization in CMOS technology which is undoubtedly the most appropriate for the required digital signal processing circuitry due to its high level of integration, low cost, and high yield.

Nevertheless, the design of an RF PA in CMOS technology has two major hurdles: first, the lower breakdown voltage due to scaling down in technology that restricts the maximum output power; and second, the reduced transconductance of MOS transistors leading to inferior power gain and poorer efficiency.

1.2.1.1 Specifications

The target frequency of operation is 5.2 GHz and the desired maximum output power is 16 dBm (40 mW)—maximum output power defined for the Unlicensed

National Information Infrastructure (U-NII) lower band [1]. Two-tone tests with the corresponding 3rd-order InterModulation Distortion (IMD3) measurements are used to assess the linearity of the PA. A limit of -35 dBc is adopted as the allowed maximum distortion. Besides the experimental characterization with 2-tone measurements, tests with OFDM signals are also performed to demonstrate the use of the dynamic supply PA in WLAN applications. The Error Vector Magnitude (EVM) of the amplified OFDM signal is measured and compared with the requirements of the regulating standard [1, 2].

1.2.2 Frequency-Tunable Capability

The objective is to develop a frequency-tunable CMOS RF power amplifier that can be used in two different frequency bands. The two first bands of choice were 2.4 and 5.2 GHz. However, because of some limitations of the circuit used to implement the proposed technique, we restricted our goal to 3.7 and 5.2 GHz. These bands are used in WLANs and are regulated by the IEEE standards 802.11y [3] and 802.11a [1]. These standards define OFDM as the modulation scheme. The metric for linearity defined by the regulating standards is the EVM. However, since we could neither simulate the EVM due to software package limitations nor measure it due to the lack of the necessary equipment,[1] we adopted a maximum IMD3 of -35 dBc as the linearity limit.

> **IMD3 Limit**
>
> It is important to justify the -35 dBc IMD3 linearity limit adopted throughout this book. It was chosen because it was more strict than the common -25 or -30 dBc found in the literature. However, we concluded afterwards that it was not strict enough if an EVM $< 3\%$ is required—refer to p. 57.

1.3 Book Outline

This book is divided in eight chapters and two appendices. When covering the two different aspects of RF power amplifiers in different chapters, some repetitions were intentionally introduced for the sake of clarity and thoroughness.

Chapter 2   Presents the basic concepts required to understand the dynamic supply PA and a survey of the main efficiency-enhancement techniques. The choice of the dynamic supply among the existing techniques is explained.

---

[1]The EVM measurements of the dynamic supply PA were carried out in Fujitsu Labs, where the necessary equipment was available. The same was not possible for the frequency-tunable PA.

Chapter 3   Focuses on the design of a dynamic supply CMOS RF PA for the 5.2 GHz band. This includes the high-efficiency, fast modulator and the RF PA itself.

Chapter 4   Shows and discusses the experimental results obtained for the dynamic supply CMOS RF PA. Measurements at 2.4 and 5.2 GHz are presented.

Chapter 5   Surveys the most recent advances in frequency-tunable capability and also establishes the basis for presenting the frequency-tunable RF power amplifier.

Chapter 6   Covers the design of the frequency-tunable CMOS RF PA.

Chapter 7   Presents the measurement results for the frequency-tunable CMOS RF PA. The difficulties encountered in the practical implementation are stressed and a discussion of the results is provided.

Chapter 8   Draws a conclusion for the two aspects treated in this book, highlighting its contributions, stressing the design and implementation difficulties and suggesting possible solutions that could be realized in future works.

Chapter 9 (Appendix A)   Explains the measurement setups used in the experimental characterization of the circuits and provides some additional screenshots for illustrative purposes.

Chapter 10 (Appendix B)   Describes the procedure that we developed and used for impedance matching of RF power amplifiers mounted on printed-circuit boards.

References

1. IEEE Standard 802.11a (1999) IEEE standard for information technology–telecommunications and information exchange between systems–local and metropolitan area networks—specific requirements, Part 11: Wireless LAN medium access control (MAC) and physical layer (PHY) specifications—high-speed physical layer in the 5 GHz band
2. IEEE Standard 802.11g (2003) IEEE standard for information technology–telecommunications and information exchange between systems–local and metropolitan area networks—specific requirements, Part 11: Wireless LAN medium access control (MAC) and physical layer (PHY) specifications—amendment 4: Further higher data rate extension in the 2.4 GHz band
3. IEEE Standard 802.11y (2008) IEEE standard for information technology–telecommunications and information exchange between systems–local and metropolitan area networks—specific requirements, Part 11: Wireless LAN medium access control (MAC) and physical layer (PHY) specifications—Amendment 3: 3650–3700 MHz operation in USA
4. Schlumpf N (2004) Adaptation dynamique de la compression d'un amplificateur RF pour des signaux modulés en amplitude et en phase. PhD thesis, EPFL, Lausanne, Switzerland, http://library.epfl.ch/theses/?nr=3020

Chapter 2
Efficiency Enhancement

Abstract This chapter presents some basic principles of RF power amplifiers, mainly dealing with the linearity–efficiency trade-off, to prepare the reader for the design of a dynamic supply CMOS RF power amplifier, subject of the next chapter. It also surveys the main efficiency-enhancement techniques found in the literature to establish a basis for comparison with the results presented in Chap. 4

2.1 Introduction

According to Kundert in [16], the two primary goals in a transmitter design are:

> First, they must transmit a specified amount of power while consuming as little power as possible. Second, they must not interfere with transceivers operating on adjacent channels.

Figure 2.1 depicts a simplified block diagram of a direct conversion transmitter. The RF power amplifier is the last component in the transmitter chain. Hence, the objectives stressed by Kundert are also valid for a PA design. These goals are better expressed through two performance metrics: efficiency and linearity. Therefore, the PA should be as efficient as possible and as linear as required by the system in which it operates. The major problem is that efficiency and linearity sit at opposite sides of a seesaw: when one goes up, the other goes down. This is the well-known linearity–efficiency trade-off inherent to RF power amplifier design.

2.2 Basic Principles

2.2.1 Linearity

The linearity of a PA is not of fundamental importance[1] in communication system transceivers with constant envelope modulation schemes. Since the RF signal

---

[1]But care should be taken to respect spurious emission limits.

P.A. Dal Fabbro, M. Kayal, *Linear CMOS RF Power Amplifiers for Wireless Applications,*   5
Analog Circuits and Signal Processing,
DOI 10.1007/978-90-481-9361-5_2, © Springer Science+Business Media B.V. 2010

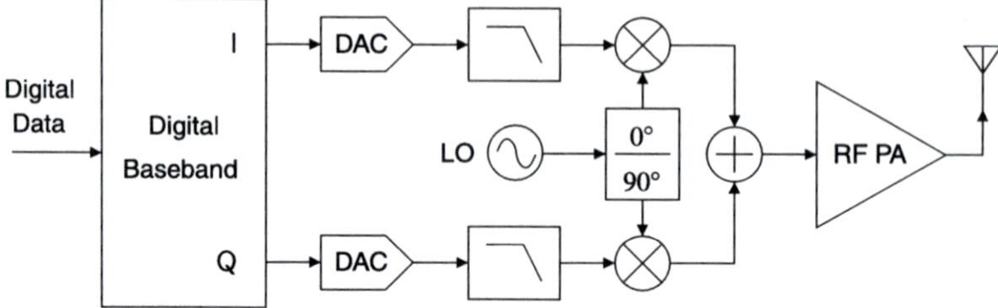

Fig. 2.1 Simplified block diagram of a direct conversion transmitter

amplitude bears no information, amplitude distortion does not alter the data to be transmitted. Constant envelope modulation schemes are used, for example, in the Advance Mobile Phone System (AMPS), Digital Enhanced Cordless Telecommunications (DECT), Global System for Mobile (GSM) communication, or Bluetooth.

On the other hand, systems employing variable envelope modulation schemes require linear RF PAs. In such a case, the RF signal amplitude contains part of the information to be transmitted. Therefore, if the RF signal undergoes amplitude distortion, the signal will not be correctly received. Variable envelope modulation schemes are used, for instance, in the North American Digital Cellular (NADC), Enhanced Data rates for the GSM Evolution (EDGE), Universal Mobile Telecommunication System (UMTS), or WLAN.

2.2.1.1 Metrics

Depending on the system, the linearity can be evaluated in a variety of ways. A linearity metric commonly used is the Adjacent Channel Power Ratio (ACPR) which measures the leakage of the RF signal into the adjacent channels. Another common linearity metric is the EVM which measures the vectorial difference between the location of the constellation points of the transmitted signal and its ideal location. A widely used method of evaluating the linearity—useful when the modulation signal cannot be generated by the available test equipment—is the 2-tone test [14, Chap. 2], [7, Chap. 9]. The linearity metric in this case is the IMD3. It is the ratio between the power of the strongest 3rd-order intermodulation product and the power of the corresponding fundamental tone. Intermodulation products of higher orders—5th or 7th—can also be used as linearity metrics as well as multi-tone tests can be performed. However, the IMD3 results of a 2-tone test are already good indicators of the linearity of a PA.

2.2.2 Efficiency

The efficiency of the PA is very important in wireless communication systems, mainly when they are employed in mobile devices. In this case, higher efficiency means longer battery lifetime which translates directly into user comfort. For non-portable applications, efficiency is also important as heat dissipation is often an issue.

2.2.2.1 Metrics

In the context of RF power amplifiers, two main efficiency metrics are commonly used: drain efficiency (η_d) and Power-Added Efficiency (PAE). The drain efficiency is calculated as the output power (P_{out}) divided by the DC power consumption (P_{DC}) as expressed below:

$$\eta_d = 100 \times \frac{P_{out}}{P_{DC}}. \tag{2.1}$$

The PAE is defined by Giannini and Leuzzi in [10, p. 164] as

> the ratio between the RF power *added* by the amplifier and the DC power required for this addition,

as shows the following equation.

$$PAE = 100 \times \frac{P_{out} - P_{in}}{P_{DC}}. \tag{2.2}$$

The PAE is a better indicator than the drain efficiency because it includes the input power (P_{in}) and, hence, the gain of the power amplifier. When the power gain (G) is higher than 10 dB, the PAE can be approximated by the drain efficiency with an error smaller than 10% as suggested by the equations below:

$$G = \frac{P_{out}}{P_{in}}, \tag{2.3}$$

$$PAE = \eta_d \left(1 - \frac{1}{G} \right). \tag{2.4}$$

2.3 Overview of the Main Efficiency-Enhancement Techniques

The linearity–efficiency trade-off is a problem that the RF PA designer has been trying to solve for a long time. Linearization as well as efficiency-enhancement techniques are possible solutions and the choice between them depends on the application and on the expertise of the designer. For application in WLAN, the requirement on the PA linearity is very stringent and, hence, it makes sense to choose to use a linear PA (class A, AB, or B) together with an efficiency-enhancement technique.

This chapter provides a review of the most important works recently published in the field of efficiency enhancement of RF power amplifiers. For more details regarding linearization techniques, the interested reader can refer to a dedicated textbook [14].

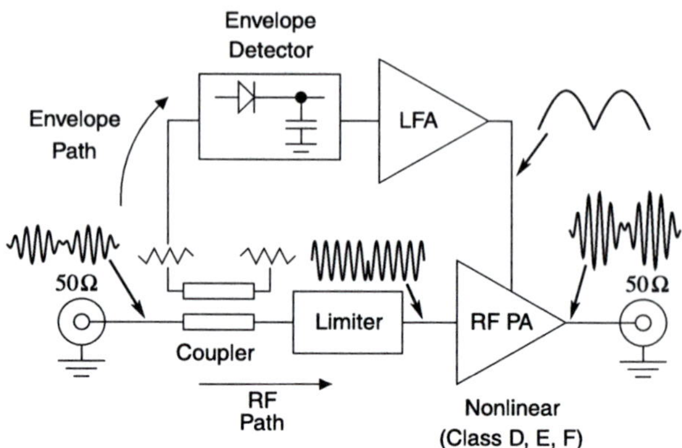

Fig. 2.2 EER technique—illustrative diagram

2.3.1 Envelope Elimination and Restoration

The Envelope Elimination and Restoration (EER) technique, also known as the Kahn technique [22], was introduced in the fifties by Kahn in [13]. Originally, Kahn developed this scheme to solve the problem of high cost and low efficiency in single-sideband[2] transmitters where typically a series of cascaded linear RF amplifiers were required. As the name suggests, in the EER technique, the *envelope* is first *eliminated* from the input RF signal, the remaining phase-modulated signal is amplified, and then, the envelope information is *restored* at the amplified output signal.

2.3.1.1 Classic Architecture

In classic EER transmitters, the envelope of the input RF signal is first detected and, then, the resulting low-frequency envelope signal is amplified by a highly-efficient, Low-Frequency Amplifier (LFA). This is the envelope path. In the RF path, the input signal passes through a limiter that removes the envelope from it. The phase-modulated signal is amplified by a highly-efficient nonlinear RF PA. The amplified envelope signal is then used to modulate the amplified output RF signal as shown in the block diagram of Fig. 2.2. Examples of classic EER systems can be found in [3, 21, 32].

2.3.1.2 Modern Architecture

In modern EER, the amplitude and phase information are generated separately in the baseband by a Digital Signal Processor (DSP). The phase-modulated signal is

---

[2]Single-sideband signals have both amplitude and phase modulation and, hence, require linear amplification.

upconverted to the RF frequency and amplified. In this case, EER becomes a misnomer since there is neither elimination nor restoration of the envelope [7, p. 311]. However, the name is still in use [17, 35].

Another name that is usually found in the literature referring to these modern EER implementations is *polar transmitter* [4, 15, 23]. This name stems from the fact that the baseband signals are converted to polar format—amplitude and phase—prior to amplification. However, this gives rise to some confusion [9, 28][3] due to the existence of a linearization technique called *polar-loop transmitter*, whose name was coined by Petrovic and Gosling in [19]. In the polar-loop transmitter, the output of the PA is sampled and converted to an Intermediate Frequency (IF) and used to linearize the power amplifier. The use of a feedback loop and baseband data in amplitude/phase domain, justifies the name of polar-loop transmitter. Examples of recent implementations of the polar-loop transmitter can be found in [1, 28].

In this book we use the term EER when referring to any system where the phase-modulated RF signal is amplified through a nonlinear amplifier and the low-frequency envelope amplitude signal is used to modulate the amplified output RF signal. In order to avoid confusion, we keep the term *polar* exclusively for polar-loop transmitters.

2.3.1.3 Advantages and Drawbacks

The main advantage of EER is the possibility of using, in variable envelope modulation systems, an efficient, nonlinear, switched-mode RF PA (class D, E, or F) instead of a low-efficiency, linear RF PA (class A, B, or AB). However, the requirements of the envelope amplifier or modulator—the output of the envelope amplifier modulates the amplitude of the output RF signal—are very stringent because it is responsible for passing the amplitude information to the output.

In order not to cancel out the benefits of the efficiency enhancement brought by the EER technique, the modulator itself must be very efficient. High-efficiency modulators are realized in practice with DC-DC converters. The stringent requirements of the EER translates into the maximum acceptable ripple at their output. A low-ripple DC-DC converter, in turn, requires a switching frequency (f_s) that is much higher than the maximum envelope bandwidth so that the low-pass filter can provide a high attenuation at f_s while passing the signals within the envelope bandwidth. However, high switching-frequency DC-DC converters are difficult to accomplish because of the losses in the switching elements. Hence, high envelope bandwidth systems can be prohibitive for the application of the EER technique.

---

[3] In [9], there is actually the elimination and restoration of the envelope. In [28], the author further separates polar transmitters in those using polar modulation prior to the PA and those using polar modulation with open-loop PA amplitude control.

2.3.1.4 CMOS IC Implementations (Refer also to Table 2.1)

Among the cited works about EER, only two are full-CMOS implementations [23, 32]. In the other works, different technologies are used. In [15], a SiGe BiCMOS technology is used to implement a class E n-FET PA and a class F Heterojunction Bipolar Transistor (HBT) PA. In [9], a 0.5 µm SiGe BiCMOS process is used to implement a bipolar variable-gain amplifier. In [3, 4], a discrete GaAs Heterojunction FET (HFET) is used to implement the PA, whereas the modulator is implemented in a CMOS technology.

Su and McFarland in [32] showed that a CMOS power amplifier using EER could reach 33% peak PAE and 3.4% EVM at an output power of 28 dBm (631 mW). The PA was designed for NADC applications, in which the signal has a Peak-to-Average Power Ratio (PAPR) of 3.5 dB [18]. The modulator attained 80% efficiency with a maximum bandwidth of 100 kHz.

Reynaert and Steyaert in [23] developed a CMOS power amplifier for EDGE transceivers in a 0.18 µm technology that attained a PAE of 22% at a maximum power of 23.8 dBm (240 mW) and 1.7% EVM (rms value). It is worth mentioning that Digital PreDistortion (DPD) and delay compensation were used in order to meet the spectral mask requirements, to improve the EVM performance, and to reestablish the spectrum symmetry.

2.3.2 Dynamic Supply RF Power Amplifier

An important drawback of the EER technique is the precision required for the amplitude modulator in replicating the input RF envelope at the supply of the PA. Another technique, called dynamic supply [8, 12, 27], solves this issue by removing the limiter from the RF path and, therefore, keeping the phase and amplitude modulation present on the signal to be amplified by the PA, as shown in the block diagram of Fig. 2.3. This, of course, requires a linear PA—which is not the case for EER—and, hence, makes the efficiency enhancement brought by this technique not as high as that achieved with EER.

In the dynamic supply technique, the power supply of the PA is varied according to the instantaneous value of the input RF signal envelope. Like EER, the envelope is extracted from the input RF signal by an envelope detector and fed to a high-efficiency modulator that transforms the low-level envelope signal into a high-level supply to the RF PA [6, Chap. 8]. However, differently from EER, in the dynamic supply PA the envelope signal is not used to modulate the PA, but only to provide a DC supply level that is just enough for the PA to amplify the input RF signal without compression [14, Chap. 8], [25]. Nevertheless, the name *modulator* for this block remains because it modulates the power supply; that is, its function is to change the amplitude of the power supply according to the instantaneous input RF signal envelope to increase the efficiency of the power amplifier.

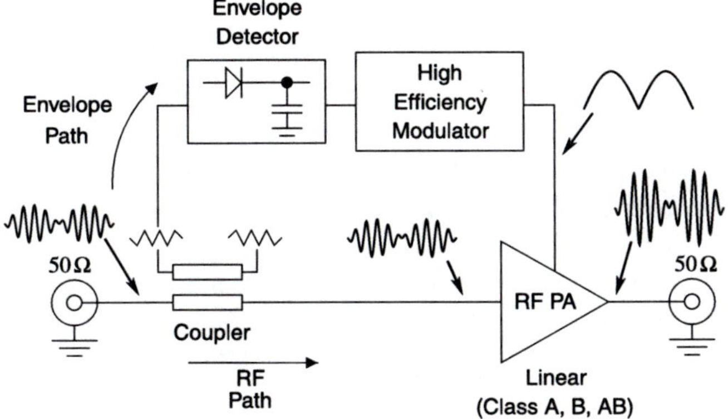

Fig. 2.3 Dynamic supply RF power amplifier block diagram

2.3.2.1 A Note on the Name of the Technique

This technique is also known under other names: *envelope tracking (ET)* [5, 17, 33], *bias adaption* [6, Chap. 8], and *envelope following (EF)* [29, 30]. However, the name *envelope tracking* is employed by Staudinger et al. in [31] to define a system in which the supply voltage of the PA is varied according to the long-term rms value of the input RF signal envelope. The name *bias adaption* might suggest that the PA operates in different classes according to the envelope amplitude, which is not the case. Although *envelope following*, used in [30, 31], could be an alternative name and a subclassification in *wideband*, *average*, and *step* for ET exists [35], in this book we use the term *dynamic supply* to avoid confusion.

2.3.2.2 Works on Dynamic Supply PAs (Refer also to Table 2.1)

Hanington et al. developed a boost converter using GaAs HBTs operating at a switching frequency of 10 MHz to improve the efficiency of a GaAs MEtal Semiconductor FET (MESFET) power amplifier operating at 4 GHz [11] and 950 MHz [12]. In [11], the supply voltage of the PA could be varied from 5 to 9 V which allowed the global efficiency of the PA to be improved by 45% (including 74% efficiency of the DC-DC converter). The PA attained an output power of 22 dBm (158 mW) with an efficiency of 43% at an IMD3 of −30 dBc. In [12], the supply voltage could be varied from 3 to 10 V, thereby improving the overall PA efficiency by 64% (74% DC-DC converter efficiency included). At 20 dBm (100 mW) output power, the efficiency was 14% and the ACPR was kept below the maximum IS-95[4] level of −26 dBc (30 kHz main channel bandwidth).

In [2, 24], the power supply voltage of a GaAs MESFET power amplifier can take two values (4 or 8 V), depending on the instantaneous value of the RF input

---

[4]IS-95—Interim Standard 95 (CDMA).

signal envelope. Instead of a DC-DC converter, a video amplifier is used to drive a bipolar transistor in an emitter-follower configuration with the envelope signal. The collector of this transistor is connected to the high-value power supply and its emitter to the drain of the GaAsFET (through an RF choke). When the envelope signal is low, the drain voltage of the GaAsFET is supplied by a low-voltage power supply through a diode. When the envelope signal reaches a certain level, the bipolar transistor conducts and supplies the voltage to the drain of the GaAsFET while turning-off the diode. With this scheme, 45% DC power consumption could be saved for operation at 4 GHz in a communication system employing Trellis Coded Modulation (TCM) with 8 dB PAPR. At 36 dBm (4 W) output power, with a 4-tone input, the intermodulation distortion was kept below -50 dBc.

Schlumpf et al. in [25–27] describe a dynamic supply PA in which the modulator is implemented in a 0.35 μm CMOS technology and the PA is realized with discrete bipolar transistors. A sliding-mode modulator controlled by hysteresis varies dynamically the supply of the PA according to the input RF signal envelope. The delay introduced by the modulator plays the role of hysteresis and together with the phase shift of the LC filter defines the switching frequency. Therefore, a small delay is desired in order to increase the switching frequency, thereby reducing the output ripple. A delay of 4 ns was achieved allowing a maximum switching frequency of 16 MHz. The system was designed for CDMA applications, for which the standard IS-95 defines a maximum Adjacent Channel Leakage Ratio (ACLR[5]) of -45 dBc (1.23 MHz main channel measurement bandwidth). The dynamic supply PA improves by 5% the PAE—a factor of 1.1 in relative terms—in comparison to the same PA operating under a constant 3.3 V power supply at the maximum linear output power (20 dBm at -45 dBc ACLR). At low output power levels, a relative improvement in efficiency of a factor of 2.1 is achieved. The rms value of the measured EVM at the maximum linear output power is 4.9% (25% peak) for the dynamic supply PA and 3.8% (19% peak) for its constant supply counterpart.

In [33], a dynamic supply PA based on discrete components is described. The modulator is implemented with discrete MOSFETs and the class AB RF PA with GaAs HFETs. The system is intended for application in WLAN transceivers operating at 2.4 GHz (IEEE 802.11g). In order to attain acceptable EVM levels, predistortion is used. An output power of 15.1 dBm (32.3 mW), a PAE of 25%, and a gain of 8 dB are attained at an EVM of 3.2%. This compares to an output power of 11 dBm (12.6 mW), a PAE of 8.2%, a gain of 10.7 dB at an EVM of 2.8% attained by the constant supply (4.4 V) power amplifier. Although the dynamic supply PA presents a much better performance than its constant supply counterpart, the comparison does not make much sense because the power amplifiers used are not the same (a MWT-871 GaAs MESFET for the constant supply and an SHF-0289 HFET for the dynamic supply PA).

---

[5]ACLR and ACPR are different names for the same linearity measurement.

Table 2.1 Summary of the main performance metrics of works on efficiency-enhanced PAs

Reference	[32]	[23]	[2]	[12]	[27]	[33]	[35]
Year	1998	2005	1995	1999	2004	2004	2006
Technique	EER	EER	Dyn.	Dyn.	Dyn.	Dyn.	Switched Dyn.
Technol. PA	CMOS	CMOS	Discr.	GaAs	Discr. Bip.	Discr. GaAs	Discr. Bip.
Technol. Mod.	CMOS	CMOS	Discr. GaAs	GaAs	CMOS	Discr. MOS	Discr. MOS
Supply (V)	3	3.3	–	3.6	3.3	6	5.5
f_c (GHz)	0.835	1.75	4	0.95	1.9	2.4	2.4
P_{out} (dBm)	28	23.8	36	20	20.5	15.1	19
PAE (%)	33	22	11.7	14	32	25	28
ACPR (dBc)	–	–	–	−26	−45	–	–
EVM (%)	3.4	1.7	–	–	4.9	3.2	2.8
IMD3 (dBc)	–	–	−50	–	−30	–	–
Mod. Eff. (%)	80	–	–	74	85	75	60
f_s (MHz)	–	–	–	10	16	7	6
Linearization	–	DPD	–	–	–	DPD	DPD
Application	NADC	EDGE	TCM	CDMA	CDMA	WLAN	WLAN

Note: Technol. = Technology; Mod. = Modulator; Eff. = Efficiency; Discr. = Discrete; Bip. = Bipolar

2.3.2.3 Dynamic Supply with Switched-Mode PA

Wang et al. in [35] call their system *hybrid EER* because the signal at the input of the PA is still a complex-modulated signal (like in the dynamic supply PA) and the PA operates in switched-mode (like in EER). However, there is neither elimination nor restoration of the envelope and, hence, the name EER for this technique is also a misnomer. As it was stated previously, the dynamic supply technique requires a linear PA so that the maximum distortion constraints can be respected. Hence, in [35], the authors used predistortion in order to circumvent the inherent distortion problems. An important contribution from their work, however, was the detailed description of the implementation of a split-band modulator. This modulator solved the problem of limited bandwidth by using a linear regulator to provide the dynamic supply when the envelope frequency is high and a buck converter to provide the dynamic supply when the envelope frequency is low. The system presented a PAE of 28% and an EVM of 2.8% at an output power of 19 dBm (80 mW). The same split-band modulator was also used in [20, 33, 34, 36].

2.4 Conclusion

This chapter presented the motivation and some basic concepts of efficiency enhancement of RF PAs. The main works dealing with PA supply modulation for efficiency enhancement purposes were covered. It was discussed that using envelope elimination and restoration can result in a high efficiency improvement, but it places a very stringent requirement on the modulator design. The dynamic supply technique, although offering lower efficiency improvement than EER, has looser modulator requirements and, hence, is a good compromise for the implementation of efficiency-enhanced PAs. Table 2.1 summarizes the main performance metrics of the most important works on efficiency enhancement found in the literature. It shows that the performance of the EER and the dynamic supply techniques are comparable and that none of the works on dynamic supply integrates both the modulator and the RF power amplifier on the same die. In the next two chapters, the design and experimental characterization of a full-CMOS implementation of the dynamic supply RF power amplifier are presented.

References

1. Akamine Y, Tanaka S, Kawabe M, Okazaki T, Shima Y, Masahiko Y, Yamamoto M, Takano R, Kimura Y (2007) A polar loop transmitter with digital interface including a loop-bandwidth calibration system. In: IEEE Int Solid-State Circuits Conf Dig Tech Pap (ISSCC'07), San Francisco, CA, pp 348–608
2. Buoli C, Abbiati A, Riccardi D (1995) Microwave power amplifier with 'envelope controlled' drain power supply In: Eur Microw Conf (EuMC'95), Bologna, Italy vol 1, pp 31–35
3. Chen JH, U-yen K, Kenney JS (2004) An envelope elimination and restoration power amplifier using a CMOS dynamic power supply circuit. In: IEEE MTT-S Int Microw Symp Dig (IMS'04), Fort Worth, TX, vol 3, pp 1519–1522
4. Chen JH, Fedorenko P, Kenney JS (2006) A low voltage W-CDMA polar transmitter with digital envelope path gain compensation. IEEE Microw Wirel Compon Lett 16(7):428–430
5. Clifton JC, Albasha L, Lawrenson A, Eaton AM (2005) Novel multimode J-pHEMT front-end architecture with power-control scheme for maximum efficiency. IEEE Trans Microw Theory Tech 53(6):2251–2258
6. Cripps SC (1999) RF Power Amplifiers for Wireless Communications, 1st edn. Artech House, Norwood
7. Cripps SC (2006) RF Power Amplifiers for Wireless Communications, 2nd edn. Artech House, Norwood
8. Dal Fabbro PA, Meinen C, Kayal M, Kobayashi K, Watanabe Y (2006) A dynamic supply CMOS RF power amplifier for 2.4 GHz and 5.2 GHz frequency bands. In: IEEE Radio Freq Integr Circuits Symp (RFIC'06), San Francisco, CA, pp 169–172
9. Elliott M, Montalvo T, Murden F, Jeffries B, Strange J, Atkinson S, Hill A, Nandipaku S, Harrebek J (2004) A polar modulator transmitter for EDGE. In: IEEE Int Solid-State Circuits Conf Dig Tech Pap (ISSCC'04), San Francisco, CA, vol 1, pp 190–522
10. Giannini F, Leuzzi G (2004) Nonlinear Microwave Circuit Design. Wiley, Chichester
11. Hanington G, Chen PF, Radisic V, Itoh T, Asbeck PM (1998) Microwave power amplifier efficiency improvement with a 10 MHz HBT. In: IEEE MTT-S Int Microw Symp Dig (IMS'98), Baltimore, MD, vol 2, pp 589–592
12. Hanington G, Pin-Fan C, Asbeck PM, Larson LE (1999) High-efficiency power amplifier using dynamic power-supply voltage for CDMA applications. IEEE Trans Microw Theory Tech 47(8):1471–1476

13. Kahn LR (1952) Single-sideband transmission by envelope elimination and restoration. Proc IRE 40(7):803–806
14. Kenington PB (2000) High-Linearity RF Amplifier Design. Artech House, Boston
15. Kitchen JD, Deligoz I, Kiaei S, Bakkaloglu B (2006) Linear RF polar modulated SiGe class E and F power amplifiers. In: IEEE Radio Freq Integr Circuits Symp (RFIC'06), San Francisco, CA, pp 475–478
16. Kundert KS (1999) Introduction to RF simulation and its application. IEEE J Solid-State Circ 34(9):1298–1319
17. Larson L, Asbeck P, Kimball D (2007) Multifunctional RF transmitters for next generation wireless transceivers. In: Proc Int Symp Circuits and Syst (ISCAS'07), New Orleans, LA, pp 753–756
18. McCune JEW (2003) Multi-mode and multi-band polar transmitter for GSM, NADC, and EDGE. In: IEEE Wirel Commun Netw Conf (WCNC'03), New Orleans, LA, vol 2, pp 812–815
19. Petrovic V, Gosling W (1979) Polar-loop transmitter. Electron Lett 15(10):286–288
20. Popp J, Lie DYC, Wang F, Kimball D, Larson L (2006) A fully-integrated highly-efficient RF class E SiGe power amplifier with an envelope-tracking technique for EDGE applications. In: IEEE Radio Wirel Symp (RWS'06), San Diego, CA, pp 231–234
21. Raab FH, Sigmon BE, Myers RG, Jackson RM (1998) L-band transmitter using Kahn EER technique. IEEE Trans Microw Theory Tech 46(12):2220–2225
22. Raab FH, Asbeck P, Cripps S, Kenington PB, Popovic ZB, Pothecary N, Sevic JF, Sokal NO (2002) Power amplifiers and transmitters for RF and microwave. IEEE Trans Microw Theory Tech 50(3):814–826
23. Reynaert P, Steyaert MSJ (2005) A 1.75-GHz polar modulated CMOS RF power amplifier for GSM-EDGE. IEEE J Solid-State Circ 40(12):2598–2608
24. Riccardi D, Abbiati A, Buoli C (1995) Linear microwave power amplifier with supply power injection controlled by the modulation envelope. International Patent, WO 95/34128
25. Schlumpf N (2004) Adaptation dynamique de la compression d'un amplificateur RF pour des signaux modulés en amplitude et en phase. PhD thesis, EPFL, Lausanne, Switzerland. URL http://library.epfl.ch/theses/?nr=3020
26. Schlumpf N, Declercq M, Dehollain C (2003) A fast modulator for dynamic supply linear RF power amplifier. In: Proc IEEE Eur Solid-State Circuits Conf (ESSCIRC'03), Estoril, Portugal, pp 429–432
27. Schlumpf N, Declercq M, Dehollain C (2004) A fast modulator for dynamic supply linear RF power amplifier. IEEE J Solid-State Circuits 39(7):1015–1025
28. Sowlati T, Rozenblit D, Pullela R, Damgaard M, McCarthy E, Koh D, Ripley D, Balteanu F, Gheorghe I (2004) Quad-band GSM/GPRS/EDGE polar loop transmitter. IEEE J Solid-State Circ 39(12):2179–2189
29. Staudinger J (2002) An overview of efficiency enhancements with application to linear handset power amplifiers. In: IEEE Radio Freq Integr Circuits Symp (RFIC'02), Seattle, WA, pp 45–48
30. Staudinger J, Gilsdorf B, Newman D, Norris G, Sadowniczak G, Sherman R, Quach T, Wang V (1999) 800 MHz power amplifier using envelope following technique. In: IEEE Radio Wirel Conf (RAWCON'99), Denver, CO, pp 301–304
31. Staudinger J, Gilsdorf B, Newman D, Norris G, Sadowniczak G, Sherman R, Quach T (2000) High efficiency CDMA RF power amplifier using dynamic envelope tracking technique. In: IEEE MTT-S Int Microw Symp Dig (IMS'00), Boston, MA, vol 2, pp 873–876
32. Su DK, McFarland WJ (1998) An IC for linearizing RF power amplifiers using envelope elimination and restoration. IEEE J Solid-State Circ 33(12):2252–2258
33. Wang F, Ojo A, Kimball D, Asbeck P, Larson L (2004) Envelope tracking power amplifier with pre-distortion linearization for WLAN 802.11g. In: IEEE MTT-S Int Microw Symp Dig (IMS'04), Fort Worth, TX, vol 3, pp 1543–1546
34. Wang F, Kimball D, Popp J, Yang A, Lie DYC, Asbeck P, Larson L (2005) Wideband envelope elimination and restoration power amplifier with high efficiency wideband envelope amplifier

for WLAN 802.11g applications. In: IEEE MTT-S Int Microw Symp Dig (IMS'05), Long Beach, CA, pp 645–648

35. Wang F, Kimball DF, Popp JD, Yang AH, Lie DY, Asbeck PM, Larson LE (2006) An improved power-added efficiency 19-dBm hybrid envelope elimination and restoration power amplifier for 802.11g WLAN applications. IEEE Trans Microw Theory Tech 54(12):4086–4099

36. Wang F, Kimball DF, Lie DY, Asbeck PM, Larson LE (2007) A monolithic high-efficiency 2.4-GHz 20-dBm SiGe BiCMOS envelope-tracking OFDM power amplifier. IEEE J Solid-State Circuits 42(6):1271–1281

Chapter 3
Design of the Dynamic Supply CMOS RF Power Amplifier

Abstract This chapter presents the design of a dynamic supply RF power amplifier for operation in the 5.2 GHz frequency band. The system includes a class A RF power amplifier and a high-efficiency, fast, switched-mode modulator used to vary the supply voltage of the amplifier. This modulator comprises a high-speed, fully-differential comparator, an anti-overlapping circuit, a synchronous switch, and an LC filter. Simulation results are presented.

3.1 Introduction

The principle of the dynamic supply RF power amplifier as an efficiency-enhancement technique was described in Chap. 2 and, in summary, it consists of varying the supply voltage of a linear PA according to the envelope of the input RF signal. For a given input power level the supply voltage must be adjusted so that it is just enough for the PA to amplify the input RF signal without compression.

Figure 3.1 shows how the power gain of a power amplifier changes with different fixed supply voltages (solid lines). With $V_{s4} > V_{s3} > V_{s2} > V_{s1}$, if the supply voltage is adjusted—at each input power level—so that the output of the PA is always close to the compression point (but below it), the dynamic supply gain curve is obtained (*dashed line*). It represents the gain of the dynamic supply PA. The compression of the output only occurs at high output power levels when the dynamic supply voltage is already adjusted to its maximum value (V_{s4}). The lowest value of the dynamic supply voltage must be limited to V_{s1} so that the gain at low power levels remains acceptable.

Based on what we argued above, the signal coming from the envelope detector must be processed prior to be applied to the input of the modulator. This envelope processing step was not shown in Fig. 2.3 on p. 11, but is now included in the new block diagram depicted in Fig. 3.2. It is responsible for amplifying the envelope signal coming from the detector, and for adding to it a floor and a ceiling, so that the PA can present a power gain close to that of Fig. 3.1 (*dashed line*).

P.A. Dal Fabbro, M. Kayal, *Linear CMOS RF Power Amplifiers for Wireless Applications*,   17
Analog Circuits and Signal Processing,
DOI 10.1007/978-90-481-9361-5_3, © Springer Science+Business Media B.V. 2010

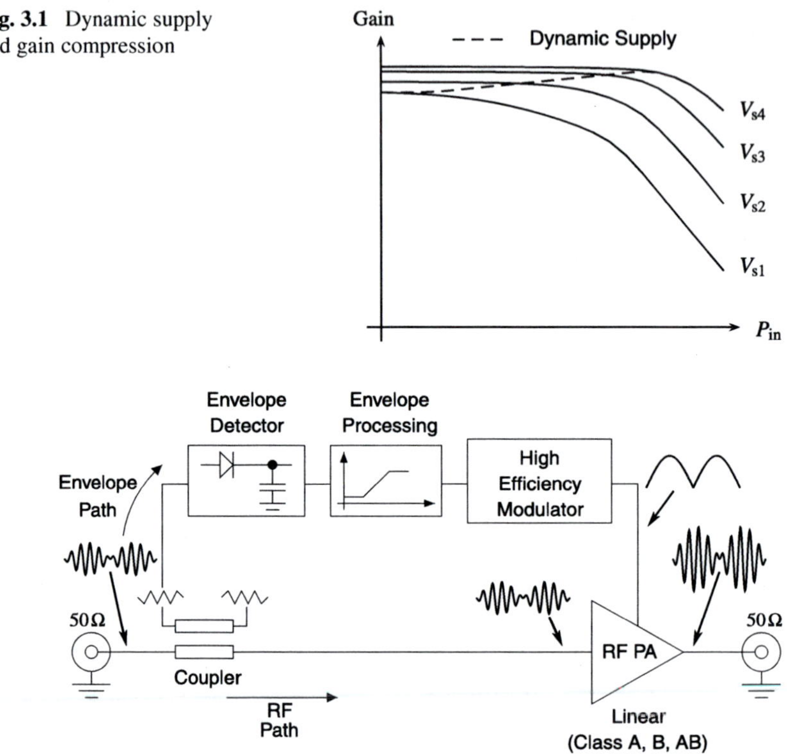

Fig. 3.1 Dynamic supply and gain compression

Fig. 3.2 Dynamic supply RF PA block diagram including the envelope processing

To describe the design of the dynamic supply PA, this chapter is divided in five sections. Two of them deal with isolated parts of the dynamic supply system. Section 3.2 presents the high-efficiency CMOS modulator design including the comparator, synchronous switch, and LC filter. Section 3.3 focuses on the design of the CMOS RF power amplifier. Section 3.4 presents the system as a whole with simulation results. Among the blocks shown in Fig. 3.2, only the modulator and the RF PA were integrated. The coupler, the envelope detector, and the envelope processing block are treated in Sect. 3.4.

3.2 High-Efficiency, Fast Modulator

As previously said, the modulator provides a voltage signal that follows the instantaneous envelope value of the input RF signal. This signal serves as the power supply for the RF power amplifier and, hence, must be able to provide any amount of current that the PA demands. In order not to lose the efficiency enhancement provided by this technique, the modulator must execute its function as efficiently as possible. This can be achieved through the use of a switched-mode power supply.

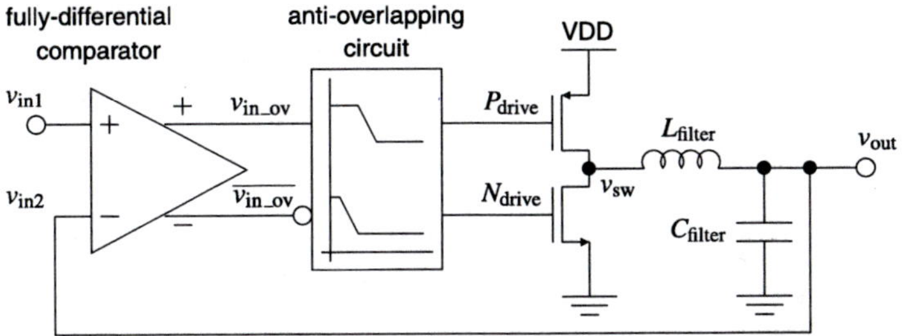

Fig. 3.3 Modulator block diagram

Figure 3.3 depicts the block diagram of the modulator which uses the same structure proposed in [28, Fig. 5.13]. It consists of a step-down DC-DC converter with a sliding-mode control. Although in Fig. 3.3 v_{out} is connected to the inverting input of the comparator, this input carries the label v_{in2} because the LC filter is placed outside the chip. This requires separate pins for v_{sw} and v_{in2}.

The sliding-mode control is a type of variable structure control where the output can have two possible states [10, Chap. 1] according to the measured value of the signal to be controlled. Due to this binary behavior, the output of the controller can directly drive the DC-DC converter. This feature promotes a very fast control action [9, Chap. 5] since it obviates the use of a device for controlling the Pulse-Width Modulation (PWM) as is the case of conventional controllers like the one used in [27]. This helps meeting the necessary speed required by the dynamic supply technique. The sliding-mode control also offers advantages over other control methods with respect to stability, robustness, and ease of implementation [5].

The switching between the output states of the modulator of Fig. 3.3 is controlled by hysteresis. The hysteresis depends on the delay (t_d) of the whole modulator chain. For the same duty cycle of the switching signal (v_{sw}), the higher is the hysteresis, the lower is the switching frequency [28, Figs. 5.7 and 5.8]. Hence, the lower is the delay, the lower is the hysteresis, and the higher is the switching frequency.

If we consider that t_d is negligible, Fig. 3.3 reveals that when the phase shift (ϕ_{LC}) introduced by the LC filter is equal to 180°, the modulator becomes unstable because this phase shift adds to the 180° of the inverting input of the comparator to which the output signal is fed back. At the frequency at which the system becomes unstable, the output of the comparator begins to oscillate. This is the switching frequency (f_s).

If we do not neglect the delay of the modulator, f_s can be calculated according to the following equation:

$$f_s t_d \times 360° + |\phi_{LC}(f_s)| = 180°. \tag{3.1}$$

It means that the total phase shift of the modulator's output signal is the sum of the phase shifts due to the delay and to the LC filter. When this sum equals 180°, the switching frequency is determined. A high switching frequency is normally desired to reduce the ripple on the modulator's output signal. Hence, the delay of the modulator must be minimized.

In the following sections, the design of the four blocks that compose the modulator is presented.

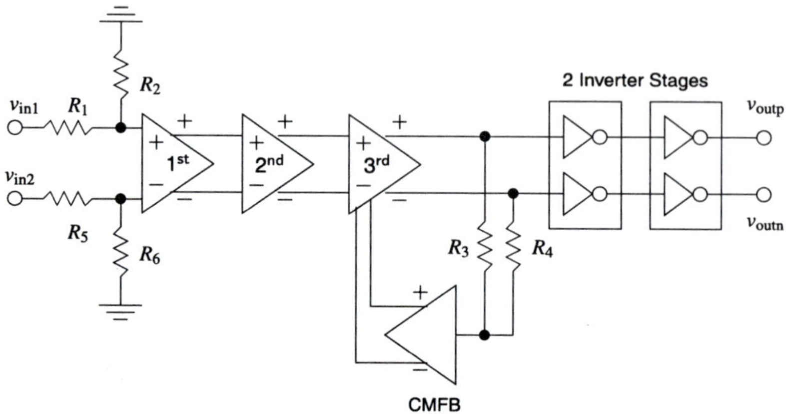

Fig. 3.4 Fully-differential comparator block diagram

3.2.1 High-Speed Comparator

The delay of the modulator chain is dominated by the comparator delay. The comparator delay can be decreased by using a high-speed comparator architecture in which a good trade-off between speed and power consumption is made. The design of high-speed comparators is well covered in [6, Chap. 8]. Figure 3.4 depicts the block diagram of the comparator designed for the dynamic supply modulator. We followed the same configuration of [28, Chap. 6], wherein further implementation details can be found. Figure 3.4 indicates that three pre-amplifying stages and two inverter stages are used. A Common-Mode FeedBack (CMFB) control is necessary because the pre-amplifying stages are fully-differential.

For small signals around the switching point, the input error signal has a higher amplitude after each amplifying stage, but after passing the three first stages the signal is still small so that we can assume linear operation at these stages. Each cascaded stage contributes to a certain propagation delay which is related to the frequency response of the stage. If we consider a single-pole frequency response, this delay can be decreased by reducing the capacitance associated to the node of highest impedance. Figure 3.5 shows the complete schematic of the comparator and these high-impedance nodes are those at the output of the three differential pairs used in the three first stages. Hence, to reduce the capacitance at these nodes we adopted cascode differential pairs with resistive loads, followed by common-drain structures before driving the subsequent stages. The purpose of using three stages is to amplify the error signal gradually so that a high gain can be achieved with a good frequency response. The gain determines the minimum required amplitude of the signal at the input so that the output of the comparator switches states. If we consider that the input signal has a finite slope, a low minimum input signal amplitude and, therefore, a high gain, decreases the delay introduced by the slope of the input signal [28, (6.6)].

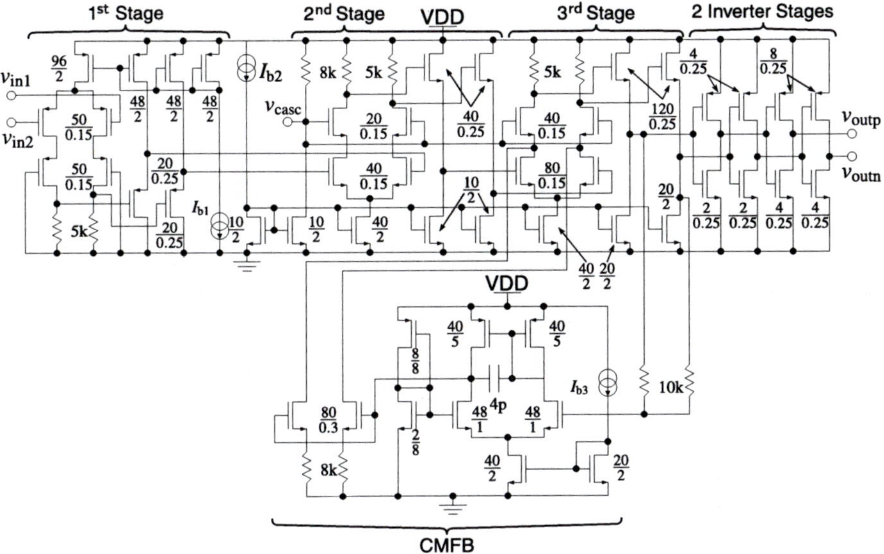

Fig. 3.5 Fully-differential comparator circuit

When the amplitude of the input error signal increases, the output of the linear stages starts slewing. Hence, a minimum size is used for the first inverter to decrease the capacitive load seen by the third pre-amplifying stage. The two inverter stages used provide a rail-to-rail output signal and a better slew rate by gradually increasing the load capacitance. Furthermore, their class B operation minimizes the current consumption.

The complete comparator circuit depicted in Fig. 3.5 includes the component dimensions. The three DC current sources ($I_{b1} = I_{b2} = 100\,\mu\text{A}$ and $I_{b3} = 5\,\mu\text{A}$) are implemented off-chip so that they can be adjusted during the characterization of the circuit. The node v_{casc} is accessible through a pad for debugging purposes.

3.2.2 Synchronous Switch

The standard structure of a buck or step-down converter is shown in Fig. 3.6(a), where V_{batt} is the supply voltage of the modulator, which in our case equals VDD. The forward voltage of the freewheeling diode when the switch S is off limits the efficiency of the converter [20, Chap. 3]. By replacing the diode D by another switch, S_2, as shown in Fig. 3.6(b), the efficiency of the converter can be increased if the switch is designed for low losses. In the case of a MOS implementation, the on resistance of the transistor, which must be minimized, is given by (3.2).

$$r_{\text{on}} = \frac{L}{\frac{\mu C_{\text{ox}}}{2} W (V_{\text{GS}} - V_{\text{T}} - V_{\text{DS}})}. \tag{3.2}$$

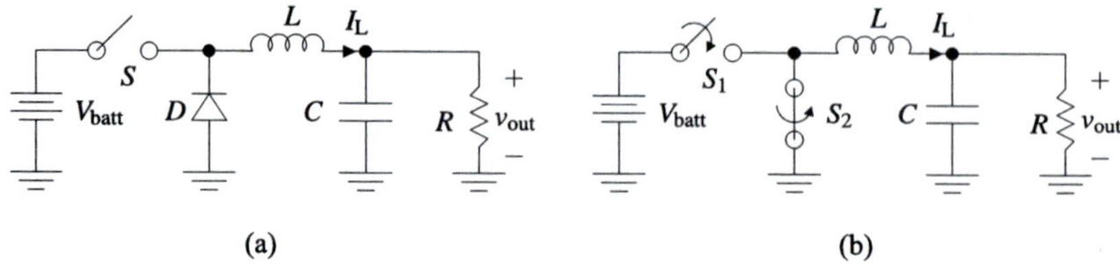

(a)   (b)

Fig. 3.6 (**a**) Standard implementation of a Buck converter with a freewheeling diode. (**b**) Implementation with a synchronous switch

Fig. 3.7 Synchronous switch

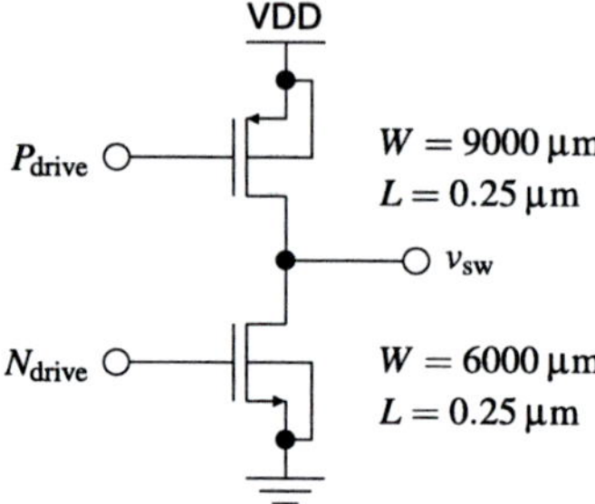

In (3.2), L and W are the transistor channel length and width, μ is the electron or hole effective mobility, C_{ox} is the gate oxide capacitance per unit area, and V_{GS}, V_{DS} and V_{T} are the gate-to-source, drain-to-source and threshold voltages. This equation reveals that a small r_{on} translates into a transistor with a very high aspect ratio [6, Chap. 4]. In order to minimize the gate capacitance and the area occupied by this big transistor, the minimum length is chosen.[1]

The dimensions of the synchronous switch are shown in Fig. 3.7, where S_1 is implemented with a pMOS transistor and S_2 with a nMOS transistor. There is, however, a trade-off between the resistive dissipation due to r_{on} and the losses due to the parasitic capacitances that must be charged and discharged in each cycle. Increasing the transistor sizes leads to a lower resistive dissipation, but also to a higher loss in the capacitance charge and discharge cycles. The expressions for the optimum values for the widths of the nMOS (W_n) and pMOS (W_p) transistors were derived in [28]:

$$\begin{cases} W_n = \dfrac{I_L}{V_{batt}} \sqrt{\dfrac{1}{(b+1)\frac{\mu_n C_{ox}}{2} C_{ox} f_s (V_{batt}-V_T)}}, \\ W_p = bW_n, \end{cases} \tag{3.3}$$

where I_L is the current through the inductor, b is the ratio between the electron and hole effective mobilities, and f_s is the switching frequency.

---

[1] The minimum length of the 0.11 μm technology used depends on the type of transistor. For 2.5 V transistors, the minimum length is 0.24 μm, whereas for 1.2 V transistors it is 0.11 μm. Although in Fig. 3.5 a 2.5 V supply voltage is used, 1.2 V devices were employed where convenient since we could guarantee that the voltage across them did not go beyond 1.2 V.

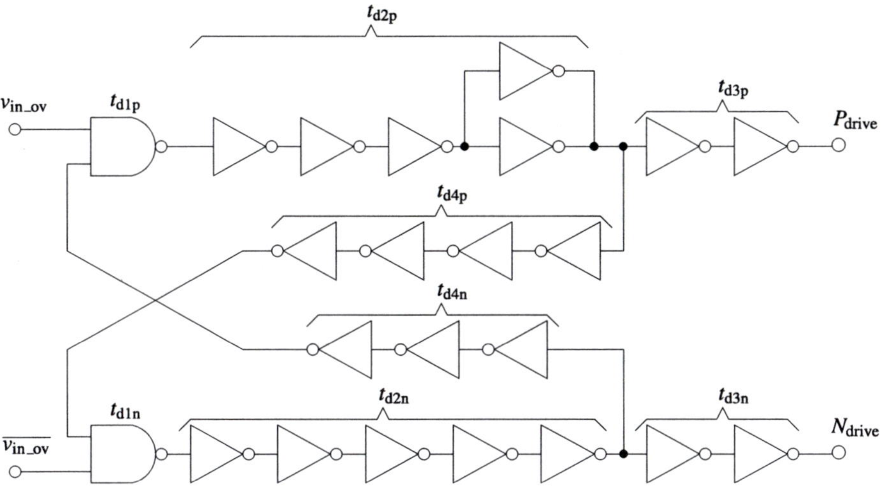

Fig. 3.8 Anti-overlapping circuit

The synchronous switch works like an inverter, but its inputs are driven separately to avoid a short circuit between VDD and ground. The two inputs are driven by signals that are delayed according to the edge of the signal. In one edge, the signal that drives the nMOS transistor is delayed relatively to the signal that drives the pMOS transistor. The opposite is true for the other edge. The anti-overlapping circuit, that is treated next, provides these signals to the synchronous switch.

3.2.2.1 Anti-overlapping Circuit

The anti-overlapping circuit is necessary to avoid that both the nMOS and pMOS transistors of the synchronous switch conduct at the same time, situation in which a short-circuit between VDD and ground would occur.

The configuration used follows that of Schlumpf in [28] and is based on the delay of an active-low RS flip-flop. Figure 3.8 depicts the circuit. Some delay cells are introduced in the RS flip-flop to better control the timing of the signals.

When $\overline{v_{in\_ov}}$ goes low, N_{drive} will have a delay of $T_{d1n} + T_{d2n} + T_{d3n}$ before going low, whereas P_{drive} will have a delay of $2 \times (T_{d1p} + T_{d2p}) + T_{d4p} + T_{d3p}$. These delays are relative to the moment when $\overline{v_{in\_ov}}$ goes low. Hence, the P_{drive} signal will be delayed compared to the N_{drive} signal by a time equal to $T_{d1p} + T_{d2p} + T_{d4p} + T_{d3p} - T_{d3n}$ in the falling edge, whereas N_{drive} will be delayed compared to P_{drive} by a time equal to $T_{d1n} + T_{d2n} + T_{d4n} + T_{d3n} - T_{d3p}$ in the rising edge. Figure 3.9 shows this timing diagram.

The two inverters in parallel in the upper path in Fig. 3.8 are used to achieve enough current capability to drive the next gates that are larger because of the larger size of the pMOS switch.

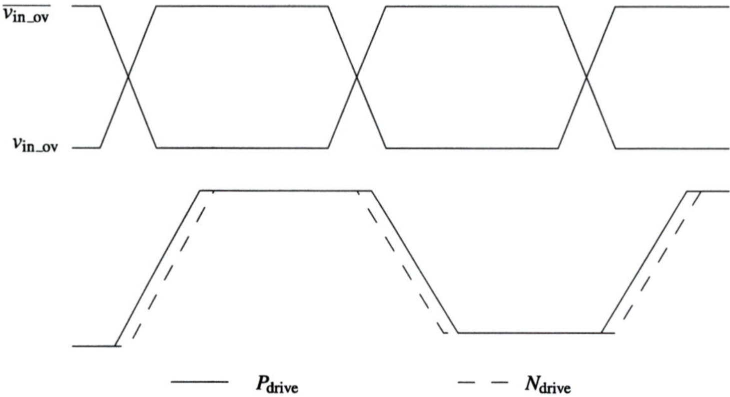

Fig. 3.9 Anti-overlapping timing diagram

Fig. 3.10 LC filter used in the output of the modulator

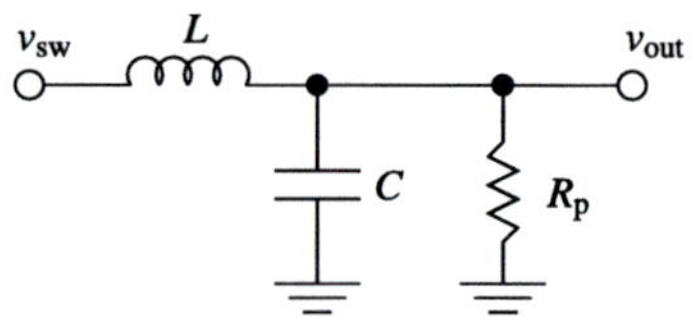

3.2.3 LC Filter

The LC filter of Fig. 3.3 attenuates the switching frequency component of the PWM signal v_{sw} and passes the envelope frequency components. In this way, the signal v_{out} will be a copy of the modulator's input signal which is the processed envelope of the input RF signal. The filter action is not perfect and, hence, the switching frequency component will be always present at the output signal. In the time domain, this is commonly referred to as ripple. For a dynamic supply RF PA, the switching frequency must be about 10 times the maximum frequency of the envelope so that the ripple in the output signal does not disturb the operation of the system [17, 28].

The LC filter, shown in detail in Fig. 3.10, has the following transfer function:

$$H_{\text{LC}}(s) = \frac{1}{1 + s\frac{L}{R_{\text{p}}} + s^2 LC},$$
(3.4)

where R_{p} is the load of the LC filter, that is, the resistance seen by the LC filter when looking at the power amplifier supply pin.

Equation (3.4) is a typical transfer function of a 2nd-order system and can be rewritten as [24, Chap. 9]:

$$H(s) = \frac{1}{1 + i2\xi\frac{\omega}{\omega_{\text{n}}} - \frac{\omega^2}{\omega_{\text{n}}^2}},$$
(3.5)

where ξ is the damping ratio, ω is the angular frequency, and ω_{n} is the undamped natural frequency.

The asymptotic frequency response of the magnitude of the transfer function of such a system is characterized by a horizontal line at 0 dB at low frequencies and by a straight line having a slope of -40 dB per decade at high frequencies. These two asymptotic lines crosses at $\omega = \omega_n$ and at the vicinity of this frequency a resonant peak occurs. The damping ratio determines the magnitude of this peak.

In the case of the LC filter for the dynamic supply PA, we are interested in a critically damped response ($\xi = \sqrt{2}/2$) for which the step response is well damped and the 3 dB cutoff frequency coincides with the undamped natural frequency. From (3.4) and (3.5) we can write ω_n as

$$\omega_n = \frac{1}{\sqrt{LC}}, \tag{3.6}$$

and the damping ratio as

$$\xi = \frac{1}{2}\sqrt{\frac{L}{C}}\frac{1}{R_p}. \tag{3.7}$$

Rearranging (3.6) and replacing into (3.7), we obtain

$$\xi = \frac{1}{2}L\omega_n\frac{1}{R_p} = \frac{1}{2}\frac{1}{C\omega_n}\frac{1}{R_p}, \tag{3.8}$$

which, rearranged, yields

$$C = \frac{1}{2\xi R_p \omega_n}. \tag{3.9}$$

Replacing (3.9) in (3.6) and rearranging it, we obtain

$$L = \frac{1}{C\omega_n^2}. \tag{3.10}$$

With (3.9), (3.10), $\xi = \sqrt{2}/2$, and the desired 3 dB cutoff frequency, the LC filter can be designed. However, as these equations depend on the load R_p, the values of L and C will be determined in Sect. 3.4 after that the simulation results for the RF PA are presented.

3.3 RF Power Amplifier

As explained in the previous chapter, the PA for a dynamic supply system must be linear. Hence, it can operate in class A, AB, or B. This section treats the design of a class A amplifier to be employed in a dynamic supply system. The reason for this choice is that an efficiency-enhancement technique has the side effect of deteriorating the linearity of the amplifier. Therefore, selecting the most linear of the classes should result in a better overall linearity performance of the whole system. Although class AB PAs can have superior linearity performance, it depends on special bias conditions [14] that may be not repeatable and, hence, must be used with

Fig. 3.11 Basic CMOS
power amplifier circuit

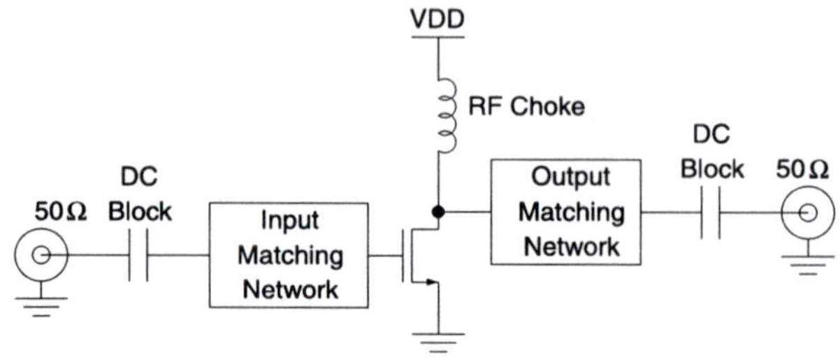

caution [12, Chap. 4]. The use of a switched-mode PA in a dynamic supply system
was reported in [29], but, in that case, the use of digital predistortion was necessary.

The basic structure of a linear CMOS RF PA is shown in Fig. 3.11. There are two
important building blocks that can be distinguished: the power amplifying stage and
the impedance matching network. Besides, an RF choke is used to provide a DC
path for the current coming from the power supply with a very high impedance at
RF frequencies. The DC block (or AC coupling) capacitors prevent the input and
output terminations of the power amplifier from disturbing its DC operating point.

Although Fig. 3.11 depicts a single-stage amplifier, 2-stage [7, 19] and 3-stage
[13, 30] PAs are common in CMOS designs due to the low power gain of a single
stage.

3.3.1 Design Procedure

The sizing of the output-stage of a linear PA depends on the maximum current (I_{max})
that the transistor must be able to provide and on two technology-dependent parameters: the breakdown (V_{BR}) and knee (V_{knee}) voltages. The maximum current is
calculated from the maximum output power (P_{out_max}) that the PA must deliver to
the load. The load is the optimum resistance (R_{opt}) obtained after the transformation
of the antenna impedance. This transformation is provided by the output matching
network.

The optimum resistance can be calculated using the loadline method described
in [12, Chap. 1]. Figure 3.12 illustrates the application of this method for a MOS
transistor with an idealized transfer characteristic. The knee voltage, as shown in
Fig. 3.12, is the turn-on voltage of the transistor [12, Chap. 2] separating the saturation region from the linear region—also known as ohmic or triode region. The
breakdown voltage is the drain voltage at which the drain current begins to increase
at a much higher rate than in the saturation region. For voltages higher than V_{BR},
a drain-substrate junction breakdown may occur [16, Chap. 1].

From Fig. 3.12, a set of equations to calculate R_{opt} can be derived. From the V_{DS}
and I_{DS} curves, the maximum output power is

$$P_{out\_max} = \frac{(V_{DC} - V_{knee}) \times I_{DC}}{2}. \tag{3.11}$$

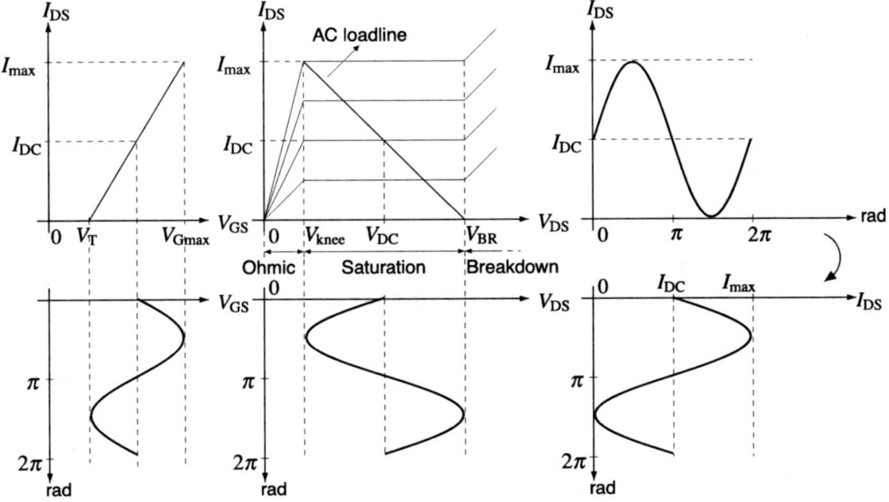

Fig. 3.12 MOS transistor's transfer characteristic and loadline for class A operation

The quiescent drain voltage (V_{DC}) and current (I_{DC}) can be written as a function of V_{BR}, V_{knee}, and I_{max}, yielding

$$V_{DC} = \frac{V_{BR} + V_{knee}}{2}, \tag{3.12}$$

and

$$I_{DC} = \frac{I_{max}}{2}. \tag{3.13}$$

Equation (3.11) can be rewritten as

$$\begin{cases} P_{out\_max} = \frac{(V_{BR} - V_{knee}) \times I_{DC}}{4}, \\ P_{out\_max} = \frac{(V_{BR} - V_{knee}) \times I_{max}}{8}, \\ P_{out\_max} = \frac{(V_{DC} - V_{knee}) \times I_{max}}{4}. \end{cases} \tag{3.14}$$

The optimum resistance is calculated by the loadline in Fig. 3.12 as

$$R_{opt} = \frac{V_{DC} - V_{knee}}{I_{DC}}, \tag{3.15}$$

which can be rewritten as a function of the maximum output power as

$$R_{opt} = \frac{(V_{DC} - V_{knee})^2}{2P_{out\_max}}, \tag{3.16}$$

or as

$$R_{opt} = \frac{(V_{BR} - V_{knee})^2}{8P_{out\_max}}. \tag{3.17}$$

The maximum output power can finally be expressed as a function of the optimum resistance as

$$P_{\text{out\_max}} = \frac{I_{\text{max}}^2 R_{\text{opt}}}{8}. \tag{3.18}$$

The equations above can be used to determine the optimum resistance and maximum drain current for a given maximum output power. Conversely, they can also be used to determine the maximum output power for given I_{max} and R_{opt}.

A first size for the transistor can be obtained from the equation for the drain current of a MOS transistor in saturation [16, Chap. 1]:

$$I_{\text{d}} = \frac{\mu C_{\text{ox}}}{2} \frac{W}{L} (V_{\text{GS}} - V_{\text{T}})^2. \tag{3.19}$$

With $I_{\text{d}} = I_{\text{max}}$ and choosing the minimum dimension for the channel length, the width of the transistor can be determined.

Figure 3.11 shows that the quiescent drain voltage of the transistor is $V_{\text{DC}} =$ VDD due to the presence of the RF choke inductor. Since the output voltage signal is symmetric with respect to VDD, the output swing is limited either by V_{knee} or by V_{BR}. Generally, in submicron CMOS technologies, the knee voltage is more restrictive than the breakdown voltage and is the limiting factor for achieving a higher output power for a given transistor size. Furthermore, V_{knee} varies not only with I_{max}, but also with the width of the transistor. If we consider that the knee voltage is more stringent than the breakdown voltage, that is, $V_{\text{DC}} - V_{\text{knee}} < V_{\text{BR}} - V_{\text{DC}}$, and that the maximum output power is known, we can use the following procedure to size the output stage of a class A power amplifier:

1. Make a first guess for V_{knee} and calculate I_{DC} using (3.11).
2. Calculate W using (3.19) and $I_{\text{d}} = I_{\text{max}} = 2I_{\text{DC}}$. V_{GS} must be chosen according to the output swing of the previous amplifying stage. If there is no constraint regarding the swing of the signal that will be available at the gate of the transistor, the choice of V_{GS} can be done based on simulation—see item 4).
3. Simulate the transfer characteristics of the transistor to check the value of the knee voltage.
4. With the new V_{knee}, return to the first point. If the value of V_{knee} is correct, calculate R_{opt} using (3.15)–(3.17).

   If V_{knee} is too high, other values for W can be simulated to find an optimum transistor size for which the maximum output power is attained with a minimum I_{max}. Care must be taken at this point because the optimum size for minimum power consumption is not necessarily the optimum size for minimum distortion. A trade-off must be made.

The core of the dynamic supply RF power amplifier is shown in Fig. 3.13. Two pads are used for the RF ground (RF GND), two for the RF input (RFIN), two for the RF output (RFOUT), and one for the gate bias (V_{bias}). An off-chip inductor that plays the role of the RF choke is also used and is connected to the RFOUT pin. The impedance matching networks are implemented with external components on

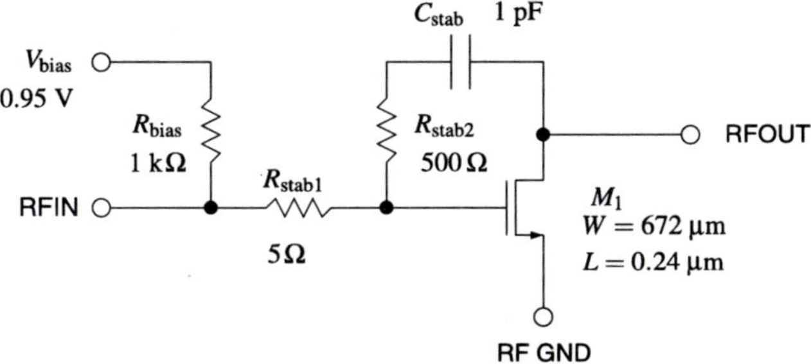

Fig. 3.13 CMOS RF power amplifier schematic (integrated parts)

the evaluation board. Following the procedure explained above, the final value of the transistor size was calculated to be $W = 672$ μm and $L = 0.24$ μm (minimum length for 2.5 V devices), with $I_{DC} = 87$ mA and for a maximum output power of 16 dBm (40 mW). The optimum resistance was calculated to be $R_{opt} = 18$ Ω.

3.3.2 Stability

The components R_{stab1}, R_{stab2}, and C_{stab} in Fig. 3.13 are used to guarantee uncon-ditional stability at all frequencies. For a circuit with only one active device, the Rollett criteria [26] can be used to determine whether the power amplifier is stable or not [8, 18]. For this purpose, the PA is considered a 2-port network.

Rollett's fundamental condition is

$$K = \frac{1 - |S_{11}|^2 - |S_{22}|^2 + |\Delta|^2}{2|S_{12}S_{21}|} > 1, \tag{3.20}$$

where K is known as the Rollett stability factor and

$$\Delta = S_{11}S_{22} - S_{12}S_{21}. \tag{3.21}$$

An auxiliary condition[2] must still be respected:

$$B_1 = 1 + |S_{11}|^2 - |S_{22}|^2 - |\Delta|^2 > 0. \tag{3.22}$$

In Agilent's simulator, Advanced Design System (ADS) [4], the conditions (3.20) and (3.22) can be verified through the functions called *stab_fact()* and *stab_meas()* [1].

These conditions stems from the fact that for unconditional stability, the real part of the impedance of port 1 and port 2 must be positive for any value of passive source or load impedances. Mathematically, this means that $|\Gamma_{IN}| < 1$ and $|\Gamma_{OUT}| < 1$ for all $|\Gamma_L| < 1$ and $|\Gamma_S| < 1$ for a given frequency range.[3] If $|\Gamma_{IN}| > 1$, the power of

---

[2] Alternative auxiliary conditions can be found in [21, 25].

[3] Γ_{IN} and Γ_{OUT} are the input and output reflection coefficients.

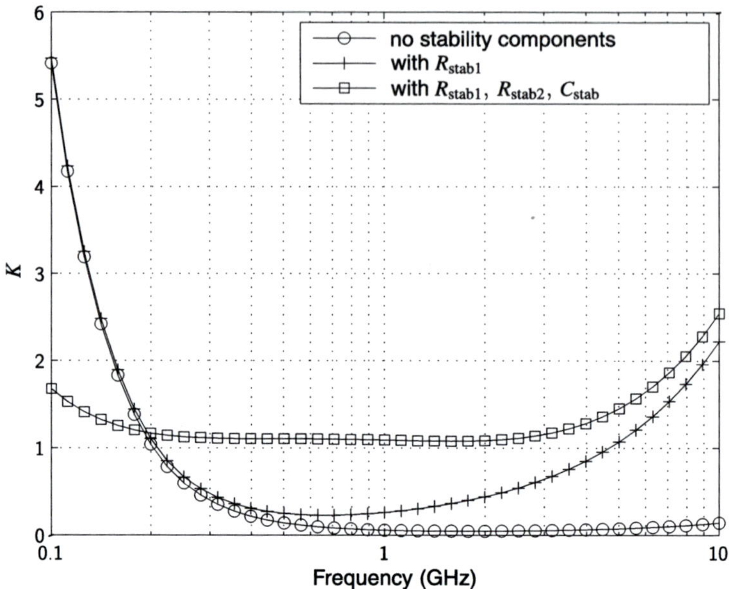

Fig. 3.14 Rollett stability factor simulation

the signal reflected in port 1 is higher than the incident signal, which, by its turn, means that the real part of the impedance at port 1 is negative. The same is valid for port 2.

However, the Rollett criteria are valid only under a proviso stated by Rollett in [26], but often overlooked:

> ... the characteristic frequencies of the twoport with ideal terminations (infinite immittances, i.e., open or short circuits, as appropriate) lie in the left half-plane.

This is known as the Rollett's proviso and, although it does not represent a problem for single active device circuits, it is important for circuits with multiple active devices [18]. Hence, the Rollett criteria can only be used to determine whether a multiple transistor PA is stable if no poles fall on the Right-Half Plane (RHP). The proviso can be verified using the Normalized Determinant Function (NDF) [18, 23].

With ADS and performing an S-parameter analysis, the factors K and B_1 were simulated and the results are shown in Figs. 3.14 and 3.15. In these figures, three cases are shown: without any stability components, with R_{stab1} only, and with R_{stab1}, R_{stab2}, and C_{stab}. Figure 3.15 reveals that the auxiliary condition $B_1 > 0$ is respected in the three cases. However, without any stability components, Fig. 3.14 shows that the condition $K > 1$ is not respected for frequencies above 200 MHz.

We begin by solving this problem with source and load stability circles [15, Chap. 3] to insure that $K > 1$ at the operating frequency of 5.2 GHz. The simulated stability circles for the "no stability components" case are shown in Fig. 3.16(a). When the load and source stability circles fall partially inside the Smith chart, the amplifier is potentially unstable. In order to push these circles

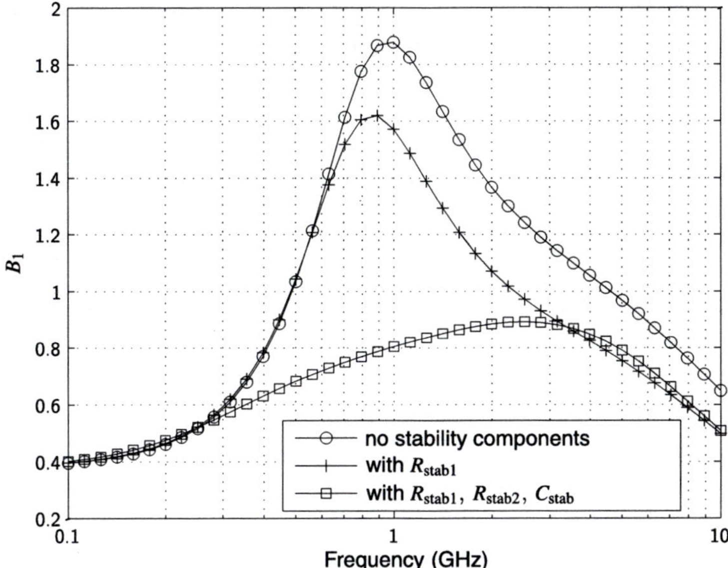

Fig. 3.15 B_1 stability factor simulation

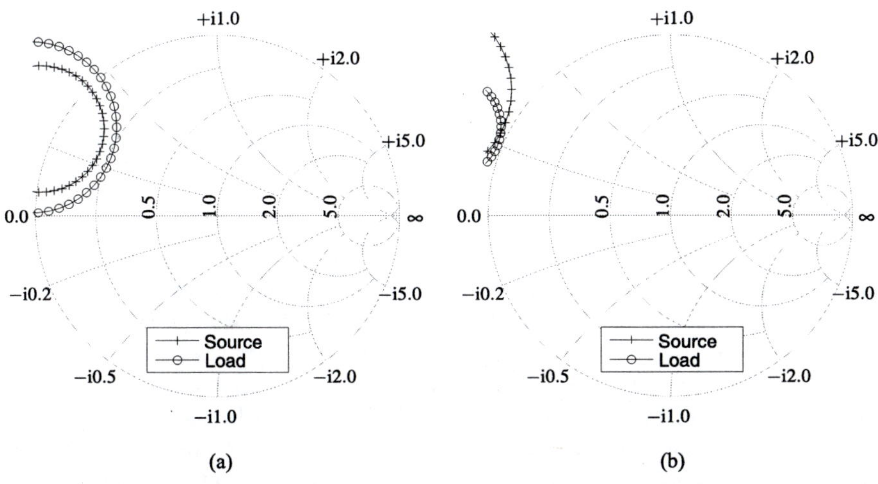

Fig. 3.16 (a) Source and load stability circles at 5.2 GHz without stability components. (b) With stability component $R_{stab1} = 5\ \Omega$

outside the Smith chart, a series resistance R_{stab1} of 5 Ω (the value of the real part of the impedance at the innermost point of the source stability circle) is placed at the input. The result shown in Fig. 3.16(b) confirms that the series resistance really makes the stability circles fall completely outside of the Smith chart. This translates into $K > 1$ at 5.2 GHz as shown by the case "with R_{stab1}" in Fig. 3.14. However, at frequencies lower than 5.2 GHz, the amplifier is potentially unstable.

A solution for ensuring stability at the lower frequencies consists of using a shunt feedback with R_{stab2} and C_{stab} from the drain of M_1 to its gate as depicted in Fig. 3.13. Due to the Miller effect, R_{stab2} and C_{stab} appear in parallel with the input and act like R_{stab1}, but for lower frequencies. C_{stab} also prevents the gate bias from being disturbed by the drain bias. The "with R_{stab1}, R_{stab2}, C_{stab}" case in Fig. 3.14 shows that $K > 1$ at all frequencies. Hence, the amplifier is stable for all possible combinations of passive source and load terminations. R_{stab1} and R_{stab2} were implemented with non-silicided polysilicon resistors, and C_{stab} with a Metal-Insulator-Metal (MIM) capacitor.

3.4 Dynamic Supply RF Power Amplifier

Besides the modulator and the PA, the dynamic supply system still requires a coupler, an envelope detector, and an envelope processing block. These components are not fundamental for the system if we consider the use of the dynamic supply PA in an modern transceiver in which the envelope is directly available from the digital baseband circuitry. However, for the system that we developed those components are necessary to make it possible to test the circuit. To simulate the dynamic supply PA in ADS, we used behavioral blocks to model these components. Experimentally, as it will be seen in Chap. 4, they were replaced by the off-chip circuitry described in that chapter.

The circuit simulated in ADS is shown in Fig. 3.17. The values of the components used are given in Table 3.1. The component FDD2P is a 2-port, frequency-domain defined, nonlinear device [2] used to detect the envelope of the input RF signal. It also performs the analog envelope processing (gain and offset) together with the component LimiterSML [3] (minimum and maximum voltage values). The resulting waveforms for the PWM and dynamic supply signals (v_{sw} and v_{out}—refer to Fig. 3.3) after a 2-tone test are depicted in Fig. 3.18. The dynamic supply voltage signal is the processed envelope with an offset and gain. It corresponds to an output power of 11.9 dBm (15.5 mW), 61 mA rms current consumption (modulator and PA), a gain of 4.9 dB, a PAE of 6.9%, and an IMD3 of -50 dBc. The minimum frequency of the switching signal v_{sw}—this frequency varies with the dynamic supply voltage level—is approximately 24 MHz. This signal is well averaged by the LC filter and the ripple observed in the supply voltage is very low (maximum 30 mV$_{pp}$). The tone spacing used in this simulation is 500 kHz and the LC filter cutoff frequency is set to 1.62 MHz.

The LC filter, which is part of the modulator, and therefore not shown in Fig. 3.17, is designed according to the impedance (Z_{PA}) seen by the modulator when looking toward the PA. This impedance, which can be represented as a resistance (R_p) in parallel with a capacitance (C_p), was simulated with ADS and the result is shown in Fig. 3.19.

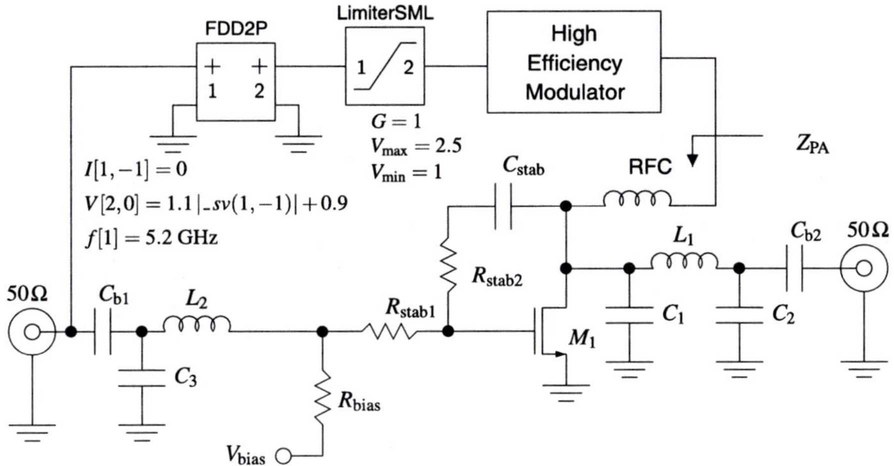

Fig. 3.17 Dynamic supply PA circuit simulated in ADS

Table 3.1 Component values used in the simulation of the dynamic supply PA

Component	Value	Component	Value
M_1	672/0.24 µm	C_2	2 pF
R_{stab1}	5 Ω	C_3	2 pF
R_{stab2}	300 Ω	C_{b1}	10 pF
R_{bias}	1000 Ω	C_{b2}	10 pF
C_{stab}	2 pF	RFC	6.8 nH
C_1	2 pF	L_2	1 nH

For the calculation of the capacitor (C_{filter}) and inductor (L_{filter}) of the LC filter, C_p is added to C_{filter} to form C in (3.9). With $\xi = \sqrt{2}/2$ and a 3 dB cutoff frequency of 1.5 MHz, C_{filter} and L_{filter} can be calculated according to (3.23) and (3.24).

$$C_{filter} = \frac{1}{2\xi R\omega_n} - C_p. \tag{3.23}$$

$$L_{filter} = \frac{1}{(C_{filter} + C_p)\omega_n^2}. \tag{3.24}$$

In order to arrive to component values that are commercially available, ω_n was changed to $2\pi \times 1.62 \times 10^6$ rad/s, resulting in $L_{filter} = 15$ µH [11] and $C_{filter} = 82$ pF [22]. The simulated transfer function of such a filter with R_p and C_p of Fig. 3.19 as load is shown in Fig. 3.20.

The switching frequency is determined according to (3.1). Hence, in Fig. 3.20, the phase (ϕ_{td}) corresponding to the delay of the modulator ($t_d = 4.7$ ns[4]) at each

---

[4] $t_d = 4.7$ ns is a simulation result. The setup for this simulation is the same used for the measurement described in Fig. 4.3 on p. 42.

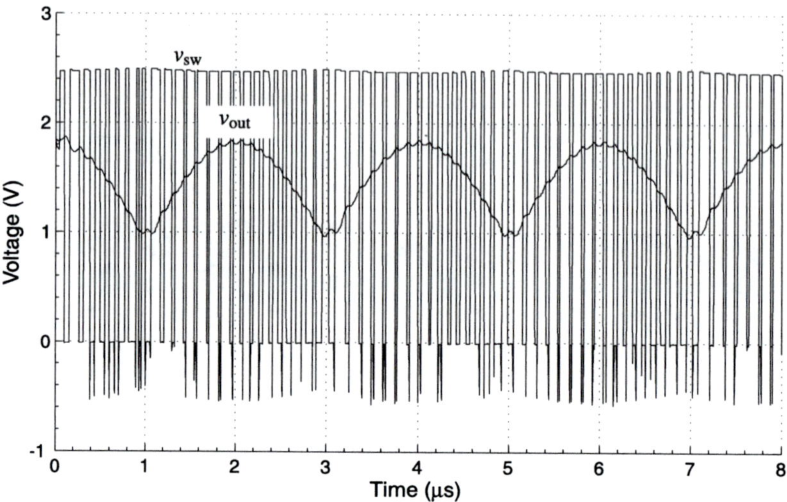

Fig. 3.18 Dynamic supply waveforms

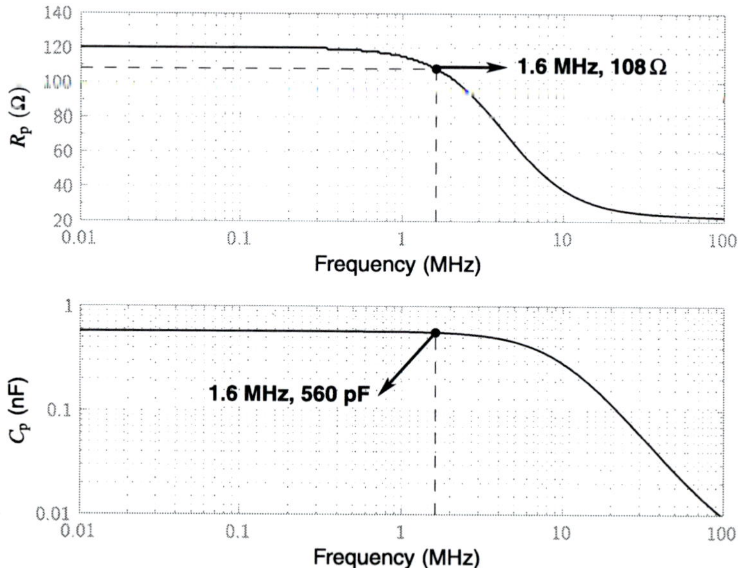

Fig. 3.19 Simulated RF power amplifier impedance

frequency must be added to the phase of the LC filter. At 33 MHz, $|\phi_{td}| = 56°$ added to $|\phi_{LC}| = 124°$ equals $|\phi_{total}| = 180°$ and, hence, the switching frequency is determined. A summary of the simulated filter characteristics is presented in Table 3.2.

Using an envelope analysis in ADS, the circuit of Fig. 3.17 was simulated with a 2-tone excitation—500 kHz tone spacing and 5.2 GHz center frequency. However,

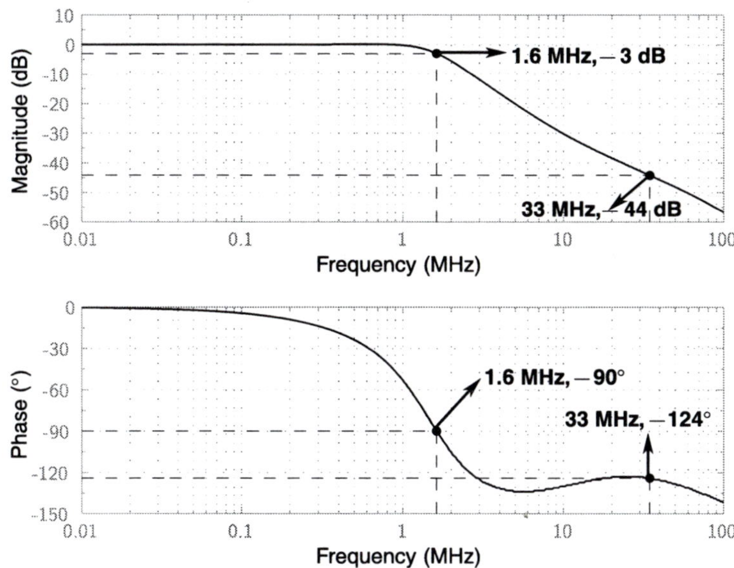

Fig. 3.20 Simulated LC filter transfer function

Table 3.2 Simulated LC filter characteristics

LC filter	at $\omega_n = 2\pi \times 1.62 \times 10^6$ rad/s	at $f_s = 33$ MHz
Attenuation	3.02 dB	43.7 dB
$|\phi_{LC}|$	89.9°	123.7°
ϕ_{td}	2.75°	56°
$|\phi_{total}|$	92.7°	179.7°
R_p	108 Ω	24.5 Ω
C_p	560 pF	51 pF

to decrease simulation time, the high-efficiency modulator circuit was replaced by a buffer with 80% efficiency. The simulation results for IMD3 and PAE are shown in Figs. 3.21 and 3.22. These figures also show the simulation results for the RF power amplifier operating with a constant VDD of 2.5 V.

Figures 3.21 and 3.22 indicate that if we adopt a maximum IMD3 level of −35 dBc as a limit, with the dynamic supply the PA reaches a higher linear output power and presents a higher efficiency than with a constant supply. More comments on these graphs are given in Sect. 4.5 on p. 50, where similar results are obtained in the experimental characterization. The simulation results at maximum linear output power are summarized in Table 3.3.

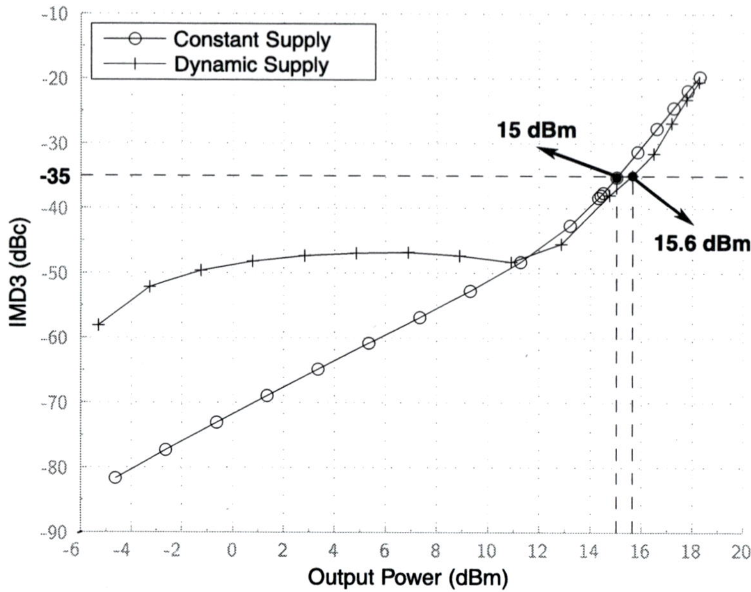

Fig. 3.21 Simulated 2-tone IMD3 at 5.2 GHz. IMD3 limit of −35 dBc is also shown (*dashed line*)

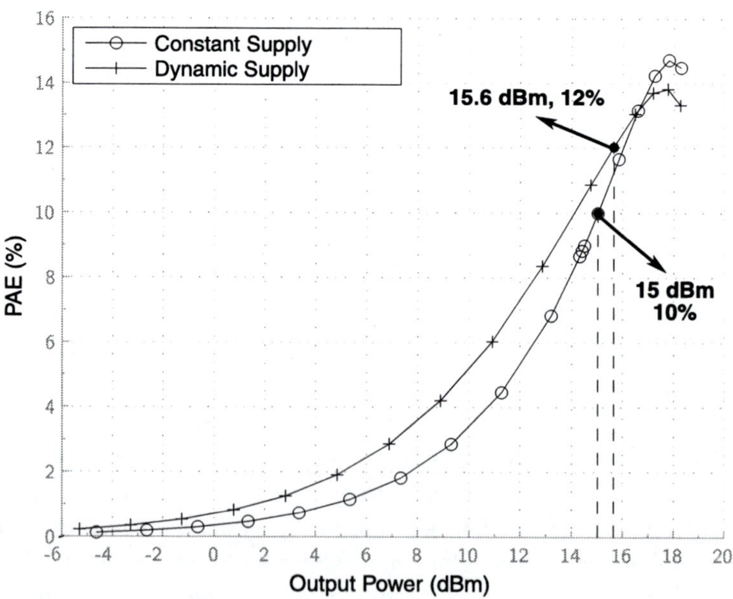

Fig. 3.22 Simulated 2-tone PAE at 5.2 GHz

Table 3.3 Two-tone simulation results at 5.2 GHz and −35 dBc IMD3

2 Tone	Constant	Dynamic
Output power (dBm)	15	15.6
PAE (%)	10	12
Gain (dB)	5	4.65

3.5 Conclusion

This chapter presented the design of a dynamic supply CMOS RF power amplifier for the 5.2 GHz frequency band. The design of the high-efficiency modulator—including the fully-differential comparator, the synchronous switch, and LC filter—and the RF power amplifier was described.

In the modulator, the main design issues are associated to the total delay of the envelope path, whereas in the PA the issues are associated to stability and impedance matching. In the next chapter, the experimental characterization of the amplifier will be presented.

References

1. ADS (2004) ADS 2004 documentation—Measurement expressions. Manual, URL http://cp. literature.agilent.com/litweb/pdf/ads2004a/expmeas/index.html
2. ADS (2004) ADS 2004 documentation—Nonlinear devices. Manual, URL http://cp. literature.agilent.com/litweb/pdf/ads2004a/ccnld/index.html
3. ADS (2004) ADS 2004 documentation—System models. Manual, URL http://cp.literature. agilent.com/litweb/pdf/ads2004a/ccsys/index.html
4. ADS (2010) Advanced Design System (ADS). URL http://eesof.tm.agilent.com/products/ads_ main.html
5. Ahmed M (2004) Sliding mode control for switched mode power supplies. PhD thesis, Lappeenranta University of Technology, Lappeenranta, Finland
6. Allen PE, Holberg DR (2002) CMOS Analog Circuit Design, 2nd edn. Oxford University Press, New York
7. Asbeck P, Fallesen C (2000) A 29 dBm 1.9 GHz class B power amplifier in a digital CMOS process. In: IEEE Int Conf Electron Circuits Syst (ICECS'00), Jounieh, Lebanon, vol 1, pp 474–477
8. Balsi M, Scotti G, Tommasino P, Trifiletti A (2006) Discussion and new proofs of the conditional stability criteria for multidevice microwave amplifiers. IEE Proc Microw Antennas Propag 153(2):177–181
9. Bühler H (1997) Réglage de systèmes d'électronique de puissance. PPUR, Lausanne
10. Büler H (1986) Réglage par mode de glissement. PPUR, Lausanne
11. Coilcraft (2005) Power chip inductors—1812PS series. Data Sheet, URL http://www. coilcraft.com/pdfs/1812ps.pdf
12. Cripps SC (2006) RF Power Amplifiers for Wireless Communications, 2nd edn. Artech House, Norwood
13. Eo Y, Lee K (2004) A fully integrated 24-dBm CMOS power amplifier for 802.11a WLAN applications. IEEE Microw Wirel Compon Lett 14(11):504–506
14. Fager C, Pedro JC, de Carvalho NB, Zirath H, Fortes F, Rosário MJ (2004) A comprehensive analysis of IMD behavior in RF CMOS power amplifiers. IEEE J Solid-State Circ 39(1): 24–34

15. Gonzalez G (1997) Microwave Transistor Amplifiers: Analysis and Design, 2nd edn. Prentice-Hall, Upper Saddle River
16. Gray PR, Hurst PJ, Lewis SH, Meyer RG (2001) Analysis and Design of Analog Integrated Circuits, 4th edn. Wiley, New York
17. Hanington G, Pin-Fan C, Asbeck PM, Larson LE (1999) High-efficiency power amplifier using dynamic power-supply voltage for CDMA applications. IEEE Trans Microw Theory Tech 47(8):1471–1476
18. Jackson RW (2006) Rollett proviso in the stability of linear microwave circuits-A tutorial. IEEE Trans Microw Theory Tech 54(3):993–1000
19. Komijani A, Natarajan A, Hajimiri A (2005) A 24-GHz, +14.5-dBm fully integrated power amplifier in 0.18 μm CMOS. IEEE J Solid-State Circ 40(9):1901–1908
20. Kularatna N (1998) Power Electronics Design Handbook: Low-Power Components and Applications. Newnes, Boston
21. Meys RP (1990) Review and discussion of stability criteria for linear 2-ports. IEEE Trans Circ Syst 37(11):1450–1452
22. Murata (2010) GRM1555C1H820JZ01—Monolithic ceramic capacitors (0402, C0G, 82pF, 50Vdc). Data Sheet, URL http://www.murata.com
23. Närhi T, Valtonen M (1997) Stability envelope-new tool for generalised stability analysis. In: IEEE MTT-S Int Microw Symp Dig (IMS'97), Denver, CO, vol 2, pp 623–626
24. Ogata K (1970) Modern Control Engineering. Prentice-Hall, Englewood Cliffs
25. Ohtomo M (1995) Proviso on the unconditional stability criteria for linear twoport. IEEE Trans Microw Theory Tech 43(5):1197–1200
26. Rollett J (1962) Stability and power-gain invariants of linear twoports. IRE Trans Circ Theory 9(1):29–32
27. Sahu B, Rincon-Mora GA (2004) A high-efficiency linear RF power amplifier with a power-tracking dynamically adaptive buck-boost supply. IEEE Trans Microw Theory Tech 52(1):112 120
28. Schlumpf N (2004) Adaptation dynamique de la compression d'un amplificateur RF pour des signaux modulés en amplitude et en phase. PhD thesis, EPFL, Lausanne, Switzerland, URL http://library.epfl.ch/theses/?nr=3020
29. Wang F, Kimball DF, Popp JD, Yang AH, Lie DY, Asbeck PM, Larson LE (2006) An improved power-added efficiency 19-dBm hybrid envelope elimination and restoration power amplifier for 802.11g WLAN applications. IEEE Trans Microw Theory Tech 54(12):4086–4099
30. Zargari M, Su DK, Yue CP, Rabii S, Weber D, Kaczynski BJ, Mehta SS, Singh K, Mendis S, Wooley BA (2002) A 5-GHz CMOS transceiver for IEEE 802.11a wireless LAN systems. IEEE J Solid-State Circ 37(12):1688–1694

Chapter 4
Measurement Results for the Dynamic Supply CMOS RF Power Amplifier

Abstract This chapter presents the experimental characterization of the dynamic supply CMOS RF power amplifier whose design was the subject of Chap. 3. The integrated circuit occupies an area of 1.35 mm^2. The circuit was designed for an operating frequency of 5.2 GHz, but measurements were also performed at 2.4 GHz. Two-tone measurement results at 2.4 and 5.2 GHz showed that the system can deliver over 16 dBm (40 mW) linear output power with efficiencies of 22.7% and 12.6%, respectively. Compared to a constant 2.5 V supply operation, the power amplifier operating with the dynamic supply delivers higher linear output power. Moreover, a relative efficiency improvement of a factor of 2.3 at 2.4 GHz and of a factor of 1.6 at 5.2 GHz at low output power levels is achieved. OFDM measurements at 2.4 GHz demonstrated that for an EVM lower than 3% and for an equal output power of 11 dBm (12.6 mW), the absolute improvement in efficiency is over 3.2%.

4.1 Introduction

We begin this chapter by describing the printed-circuit board designed for the evaluation of the circuits. This board is necessary because the dynamic supply RF power amplifier requires some circuit components that were not integrated, namely, the input and output impedance matching networks, three bias current sources, the LC filter at the output of the modulator, and decoupling capacitors. Then we proceed to the characterization of the high-speed, high-efficiency modulator in Sect. 4.3 and the PA in Sect. 4.4. Once these two fundamental blocks are characterized, we present the measurement results of the complete dynamic supply system in Sect. 4.5.

4.2 Main Evaluation Board

A Printed-Circuit Board (PCB) was designed and fabricated for the evaluation of the dynamic supply RF power amplifier. A special RF laminate, Rogers RO4350 [25],

P.A. Dal Fabbro, M. Kayal, *Linear CMOS RF Power Amplifiers for Wireless Applications,*   39
Analog Circuits and Signal Processing,
DOI 10.1007/978-90-481-9361-5_4, © Springer Science+Business Media B.V. 2010

Table 4.1 RO4350 characteristics [25]

Parameter	Value
Thickness (mm)	0.508
ε_r	3.48
$\tan\delta$	0.0031 @ 2.5 GHz
Copper cladding (µm)	17

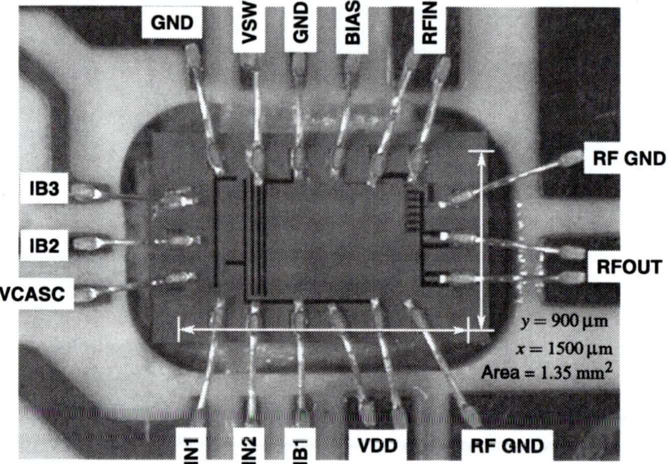

Fig. 4.1 Photograph of the dynamic supply die lodged in the evaluation PCB

was used as the substrate. The reason for choosing an RF laminate, instead of a conventional FR-4,[1] stems from its superior RF performance, mainly the tight tolerances specified for its electrical characteristics, such as the relative dielectric constant (ε_r), and its low dielectric loss ($\tan\delta$ [24, Chap. 1]). The main characteristics of the substrate used are given in Table 4.1.

The 2-layer PCB was fabricated with a special window opened in it where the die of the dynamic supply RF PA was inserted. The photograph of the assembly in Fig. 4.1 shows the window used to reduce the length of the bondwires resulting from the chip-on-board mounting. The equivalent inductances from the pad of the integrated circuit to the pad of the PCB were therefore reduced. The bottom layer of the PCB was used as a ground plane and several metalized vias were employed to connect the ground nodes in the top layer to the ground plane. The back of the die was also connected to the ground plane using a piece of metal sheet and conductive glue. With such an arrangement, bondwires with a length of less than 0.5 mm were possible in critical points like the RF input, RF output and RF ground, resulting in less than 0.5 nanohenry per bondwire—if we consider a value of 1 nH/mm [15,

---

[1]FR-4 stands for Flame Retardant type 4, which is a woven fiberglass reinforced epoxy resin.

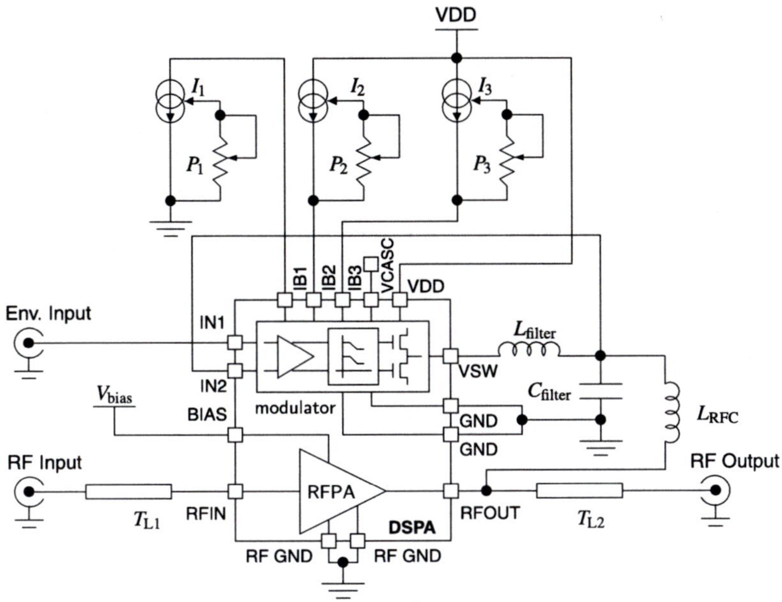

Fig. 4.2 Dynamic supply PA (DSPA) evaluation board schematic diagram

p. 53]. In the photograph in Fig. 4.1, the pin names are also shown so that the circuit nodes with similar names (refer to Fig. 3.4 on p. 20, Fig. 3.5 on p. 21, and Fig. 3.13 on p. 29) can be recognized. The total die area including pads is 1.35 mm^2.

The evaluation board schematic is shown in Fig. 4.2. The dynamic supply PA IC includes the modulator and the power amplifier. Three adjustable current sources [22] were used to bias the modulator. Their nominal values are $I_1 = I_2 = 100$ µA and $I_3 = 5$ µA. They can be adjusted with the potentiometers P_1, P_2, and P_3. The other external components are the inductor and capacitor that form the LC filter, the flange-mount SMA connectors [10] used for the RF input and output, and the SMB connector [11] for the envelope input. T_{L1} and T_{L2} are identical microstrip transmission lines used in the input and output. Their width and length are 1100 µm and 6200 µm, providing a characteristic impedance of $Z_0 = 50$ Ω. Decoupling capacitors were also used in the supply rails and voltage bias nodes, but are not shown in the figure. The measurement setup used in the characterization can be found in Chap. 9 (Appendix A). The microstrip transmission lines at the input and output influence the impedance matching. A procedure that helped matching the RF power amplifier mounted on a printed-circuit board is described in Chap. 10 (Appendix B).

The main evaluation board whose schematic diagram is shown in Fig. 4.2 does not include the envelope detector and processing blocks of Fig. 3.2 on p. 18. These blocks were implemented in another PCB described in Sect. 4.3.1, whose output is connected to the "Env. Input" of the main evaluation board.

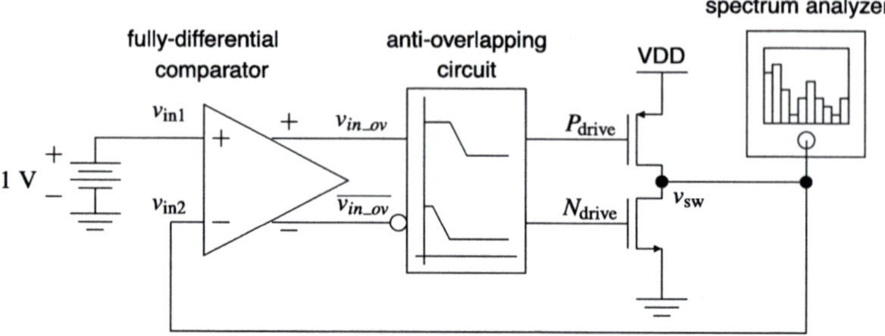

Fig. 4.3 Setup used in the measurement of the modulator delay

4.3 Modulator Characterization

Besides verifying that the modulator executes its function correctly, the most important parameter to be measured is its delay (t_d). As it was seen in the last chapter, this delay and the phase shift of the LC filter determine the maximum switching frequency according to (3.1), repeated here for convenience:

$$f_s t_d \times 360° + \phi_{LC}(f_s) = 180°. \tag{4.1}$$

In order to measure the modulator delay, the setup depicted in Fig. 4.3 was used. A constant voltage of 1 V was applied to one of the inputs (v_{in1}) of the modulator and its output (v_{sw}) was connected to the other input. The self-oscillation frequency was measured as if the modulator were operating as a ring oscillator. In this case, the output v_{sw} oscillates at the maximum frequency as if $\phi_{LC}(f_s) = 0°$ in (4.1). Figure 4.4 shows the screenshot of the spectrum analyzer[2] from which we identify the maximum switching frequency to be 112.4 MHz.

Replacing $\phi_{LC}(f_s) = 0°$ and $f_s = 112.4$ MHz in (4.1), the resulting delay is $t_d = 4.5$ ns.[3]

4.3.1 Envelope Detection and Processing

In Fig. 3.1 on p. 18, it was shown that the dynamic supply voltage must be a function of the input power and must respect some conditions. There is a minimum supply voltage that is determined by the minimum power gain desired and the knee voltage of the power transistor. There is also a maximum voltage that cannot exceed VDD.

---

[2] Agilent's HP8562E, 30 Hz–13.2 GHz [2].

[3] In [7], we published $t_d = 8.9$ ns, due to a mathematical mistake. The correct is $t_d = 4.5$ ns. This mistake, however, did not affect the results presented in that paper.

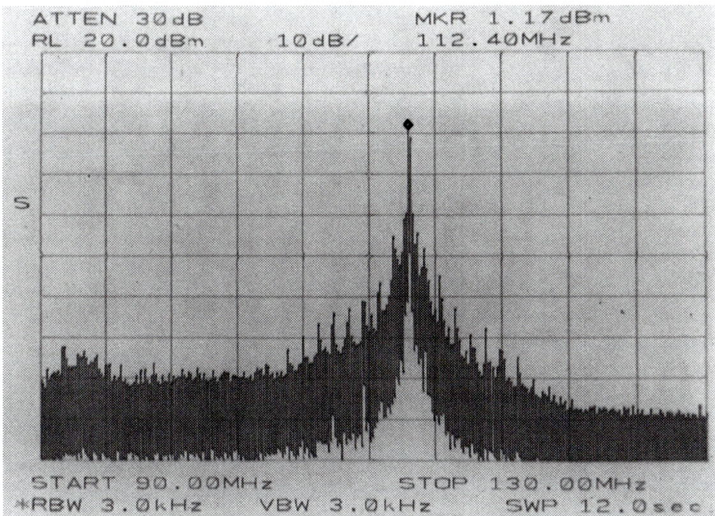

Fig. 4.4 Spectrum analyzer screenshot showing the maximum switching frequency

Fig. 4.5 Envelope processing function

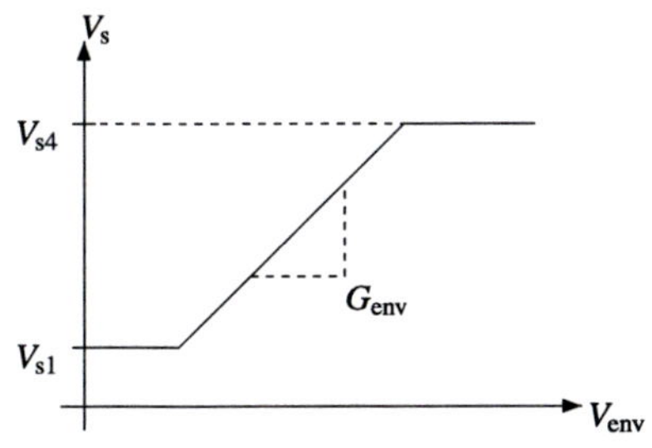

An example of a function, shown graphically in Fig. 4.5, that complies with these requirements is

$$V_s(P_{in}) = \begin{cases} V_{s1} & \text{if } G_{env} V_{env} + V_{knee} < V_{s1}, \\ G_{env} V_{env} + V_{knee} & \text{if } V_{s1} \leq G_{env} V_{env} + V_{knee} \leq V_{s4}, \\ V_{s4} & \text{if } G_{env} V_{env} + V_{knee} > V_{s4}, \end{cases} \qquad (4.2)$$

where G_{env} is the envelope amplification gain, V_{env} the input envelope amplitude, and V_s the dynamic supply voltage which is a function of the input power (P_{in})—with V_{s1} and V_{s4} being the inferior and superior limits as suggested in Fig. 3.1 on p. 18.

The circuit that implements the function described in (4.2) and illustrated in Fig. 4.5 is shown in Fig. 4.6. The main component values are given in Table 4.2. The commercial power detector LTC5508 [17] from Linear Technology is used at the input for envelope detection. The detected envelope (v_{env}) is then applied to the input of the envelope amplifier. The gain, and the minimum and maximum dynamic supply voltages can be adjusted. The core of this amplifier is the wideband operational amplifier OPA690 [5] from Burr-Brown. This circuit was implemented in a

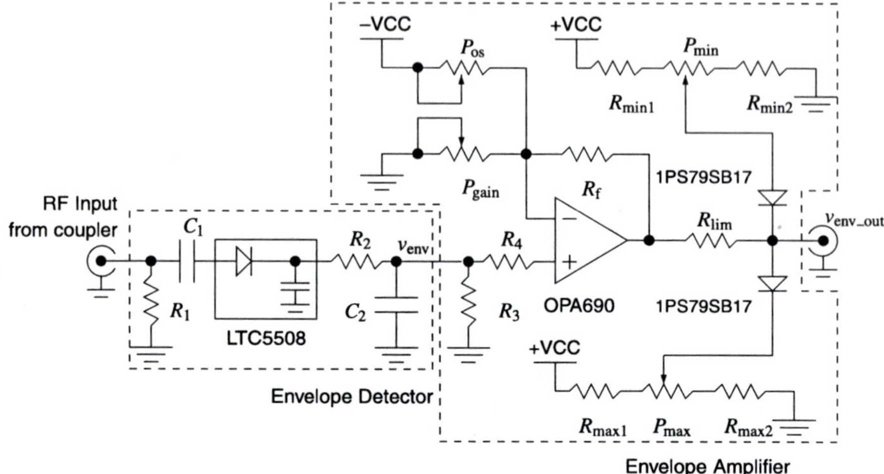

Fig. 4.6 Envelope processing circuit including envelope detection

Table 4.2 Envelope processing circuit details

Component	Value
R_1	68 Ω
R_2	68 Ω
R_3	50 Ω
R_4	175 Ω
R_f	330 Ω
C_1	18 pF
C_2	100 pF
VCC	5 V

PCB and used together with the main evaluation board in the characterization of the dynamic supply RF PA.

In the detector of Fig. 4.6, R_1 is a termination resistor[4] and C_1 is a coupling capacitor [16]. R_2 and C_2 form a low-pass output filter with a cutoff frequency of around 20 MHz. A single power supply of 5 V is used for the LTC5508 power detector.

In the envelope amplifier of Fig. 4.6, a dual power supply of ± 5 V is used. R_3 is a termination resistor and R_4 is an input bias current canceling resistance for the OPA690 operational amplifier, as indicated in its datasheet [5]. R_f is selected to be between the values suggested by the operational amplifier's manufacturer ($200\ \Omega < R_f < 1.5$ kΩ and $R_f // P_{gain} < 300\ \Omega$) while being able to provide a gain

---

[4]68 Ω in parallel with the 250 Ω internal termination of the LTC5508 results in a termination of about 50 Ω seen by the coupler.

selection from one to four through the potentiometer P_{gain}. Three other potentiometers are used in the envelope amplifier: P_{os} for adjusting the voltage offset of the output envelope signal, and P_{min} and P_{max} (together with the 1PS79SB17 Schottky diodes [23]) for settling its minimum and maximum voltage values [14]. Resistor R_{lim} is used to limit the current flowing through the diodes when the minimum and maximum voltages are exceeded.

4.4 RF Power Amplifier Characterization

The RF power amplifier was characterized using the main evaluation board designed for the dynamic supply PA. The dynamic supply circuit is disabled and the supply voltage of the PA is set to VDD = 2.5 V. A 0402 case-size chip inductor of 6.8 nH [21] was used as the RF choke (L_{RFC} in Fig. 4.2). Its impedance is about 800 Ω at 5.2 GHz [18], which is much higher than the 18 Ω optimum resistance, as required.

4.4.1 S-Parameters Measurement

The 2-port S-parameter measurements were made with the network analyzer HP8719D [1] calibrated from 1.7 to 8.7 GHz with the Short-Open-Load-Thru (SOLT) calibration kit [3] provided by the manufacturer.

First, the RF PA was characterized without impedance matching networks. The measured input (S_{11}) and output (S_{22}) reflection coefficients are shown in the Smith chart of Fig. 4.7, where it can be seen that the PA is totally unmatched at 5.2 GHz. The forward (S_{21}) and reverse (S_{12}) power gains are shown in Fig. 4.8. At 5.2 GHz, the forward power gain is only 0.3 dB confirming that the amplifier is unmatched.

Next, using the procedure described in Chap. 10 (Appendix B), the power amplifier was matched at 5.2 GHz. The resulting S parameters are shown in Figs. 4.9 and 4.10. In this matched condition—it can be seen that at 5.2 GHz both S_{11} and S_{22} are near the center of the Smith chart—the power amplifier attains a power gain of 4.6 dB. Although this gain is low, low power gains are relatively common in single-stage CMOS PA designs [8].

Another evaluation board was also prepared with the same power amplifier, but with the impedance matching made for 2.4 GHz. In this case, a 10 nH inductor [20] was used as the RF choke, presenting an impedance of about 180 Ω at 2.4 GHz. According to the procedure described in Chap. 10 (Appendix B), the RF PA was matched at 2.4 GHz and the resulting S parameters are shown in Figs. 4.11 and 4.12. The power amplifier reaches a power gain of 8.4 dB. A summary of these S-parameter results for 2.4 and 5.2 GHz are given in Table 4.3.

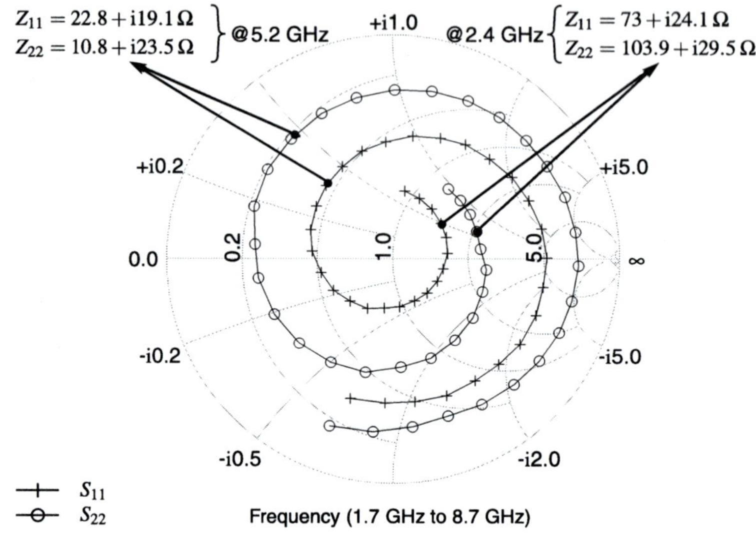

Fig. 4.7 Unmatched power amplifier SOLT measurement (S_{11} and S_{22})

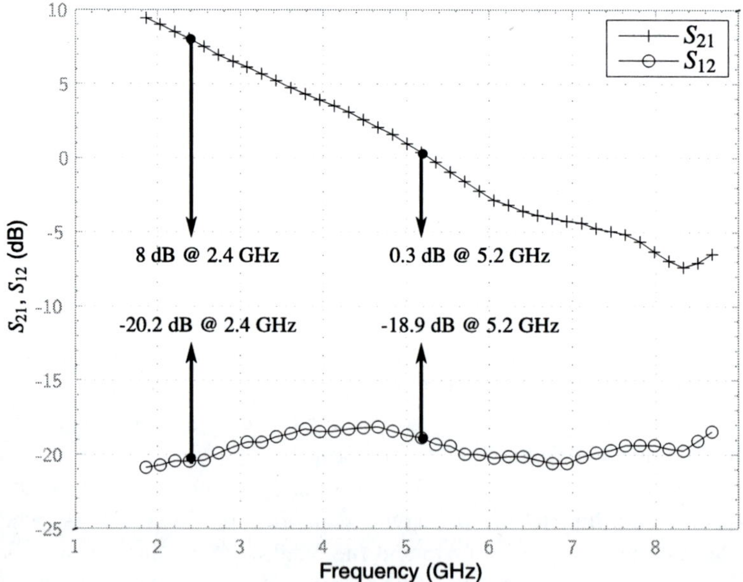

Fig. 4.8 Unmatched power amplifier SOLT measurement (S_{21} and S_{12})

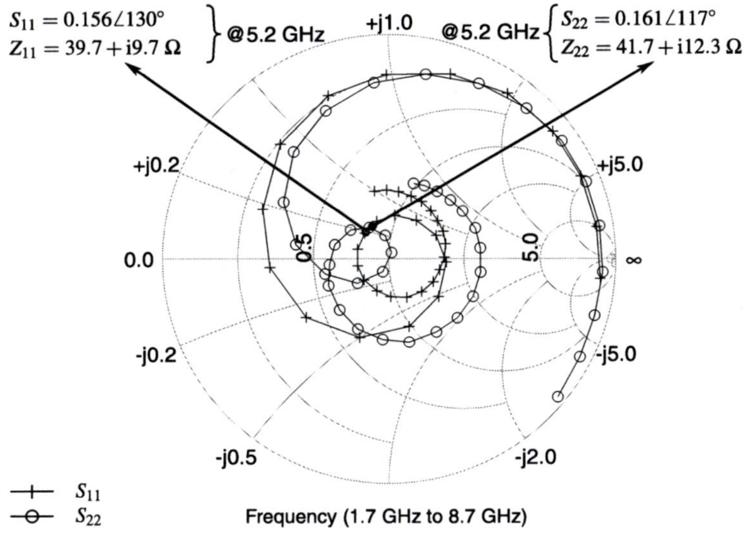

Fig. 4.9 SOLT measurement of the PA matched at 5.2 GHz (S_{11} and S_{22})

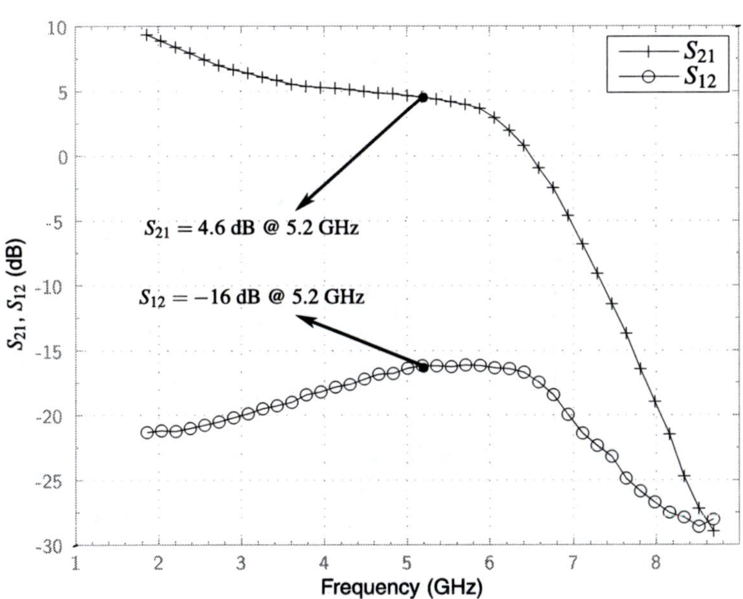

Fig. 4.10 SOLT measurement of the PA matched at 5.2 GHz (S_{21} and S_{12})

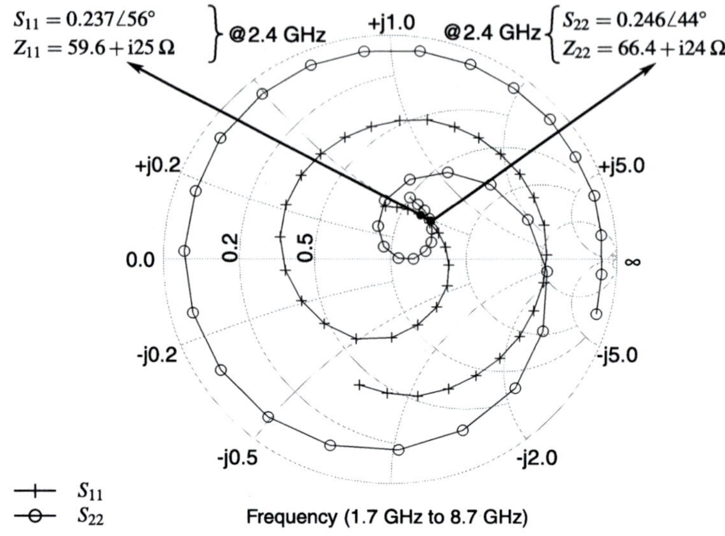

Fig. 4.11 SOLT measurement of the PA matched at 2.4 GHz (S_{11} and S_{22})

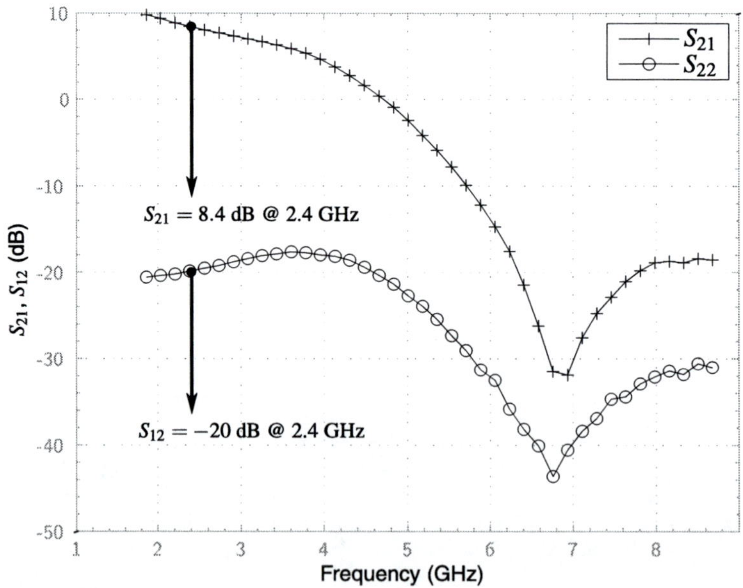

Fig. 4.12 SOLT measurement of the PA matched at 2.4 GHz (S_{21} and S_{12})

Table 4.3 Summary of measured S parameters at matched conditions

SP	2.4 GHz	5.2 GHz
S_{11} (dB)	−12.6	−16.2
S_{21} (dB)	8.4	4.6
S_{22} (dB)	−12.2	−15.9
S_{12} (dB)	−19.9	−16

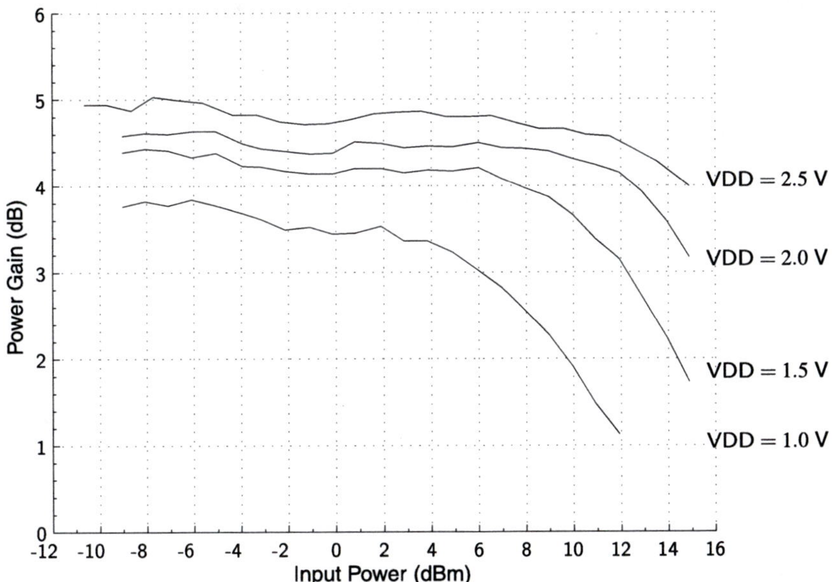

Fig. 4.13 Measured CW power gain at 5.2 GHz

4.4.2 Single-Tone Measurements

Single-tone, or Constant Wave (CW), measurements can be used to assess the linearity of a PA in terms of its 1 dB compression point. We have preferred, however, to use 2-tone tests and be able to assess quantitatively the nonlinearity level at each output power level through the measurement of the IMD3. The 2-tone measurement results showing the efficiency and linearity behavior of the power amplifier are treated in the next section where we compare the results for a constant 2.5 V supply operation with those for the dynamic supply operation. We present, nevertheless, in Fig. 4.13, the single-tone power gain of the PA at 5.2 GHz for different fixed supply voltages. This is done so that a comparison among the S-parameter, single-tone, and 2-tone power gains can made in Sect. 4.5.1 on p. 50.

4.5 Dynamic Supply RF PA Characterization

The dynamic supply RF power amplifier was characterized using the setup described in Chap. 9 (Appendix A). It requires two evaluation PCBs: the board to which we referred as the main evaluation board and an auxiliary board with the envelope detection and processing circuits. The RF signal is split in two, using a power splitter, with one signal going to the PA and the other to the envelope detector. Two-tone measurements were used to verify the linearity and efficiency of the power amplifier operating in the dynamic supply system and with a constant supply voltage. Furthermore, measurements with an OFDM signal were also performed to evaluate the system for application in WLANs.

4.5.1 Two-Tone Measurements

The 2-tone measurements were carried out with tones centered at 5.2 GHz and spaced 500 kHz apart from each other. We used the RF signal generator SMIQ06B [27] and the RF signal analyzer FSIQ7 [26]. Detailed information on the setup used for this measurement are provided in Chap. 9 (Appendix A).

Figure 4.14 shows the measured power gain. Five curves are presented with four of them for different cases of the constant supply operation, that is, 1 V to 2.5 V in 0.5 V steps. These curves were used to calibrate the envelope processing circuit to set the dynamic supply voltage to a level just enough to result in amplification without compression. The fifth curve (*thick line*) shows the dynamic supply PA power

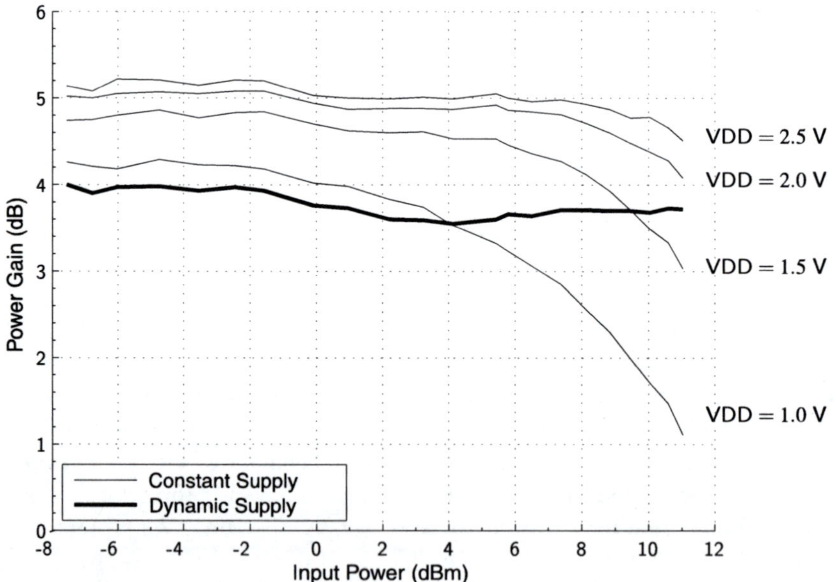

Fig. 4.14 Measured 2-tone power gain at 5.2 GHz

gain, which is almost constant. Its flatness suggests that the dynamic supply PA should present an extended linear range in comparison with the constant 2.5 V supply operation.

If we compare Figs. 4.10, 4.13, and 4.14, we can notice a difference between the different power gain measurements. The S-parameter power gain is 4.6 dB at 5.2 GHz (measured with an input power of -25 dBm), whereas the single-tone and 2-tone power gains are close to 5 dB. The explanation for this difference is that for the single- and 2-tone measurements the instruments used for the test are a signal generator and a spectrum analyzer, whereas for the S-parameter it is a network analyzer.[5] Hence, in changing from one setup to the other, some measurement errors may have been introduced.

> In the figures that will follow, the results for the dynamic supply RF power amplifier are always compared to those for the same RF PA operating at a constant supply voltage of VDD $= 2.5$ V.

Figure 4.15 depicts the IMD3 as a function of the output power. For low output power levels, the dynamic supply PA generates higher distortion than with a constant

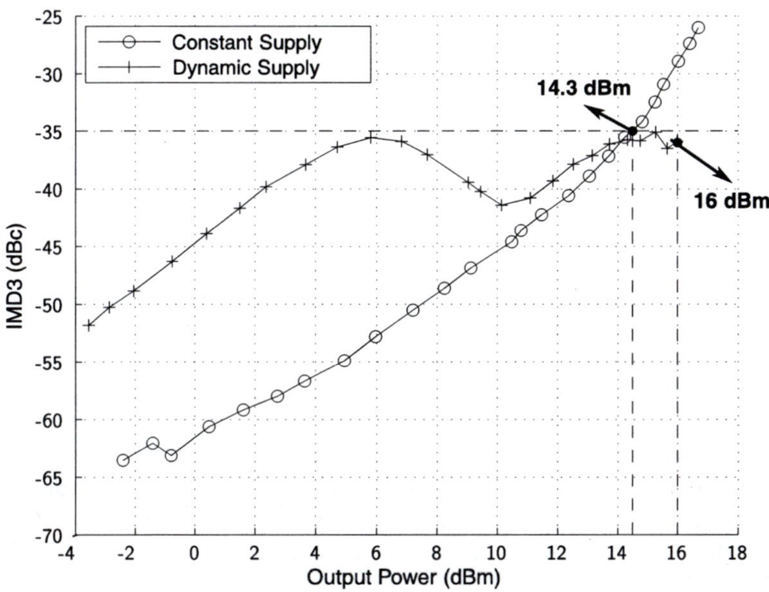

Fig. 4.15 Measured 2-tone IMD3 at 5.2 GHz. IMD3 limit of -35 dBc is also shown (*dashed line*)

---

[5] A sweep in the input power from -10 to 10 dBm was made with the network analyzer at 5.2 GHz. It revealed that the value of S_{21} was also approximately 4.6 dB from -10 to 0 dBm. This value decreased for higher input power levels as the amplifier began to compress.

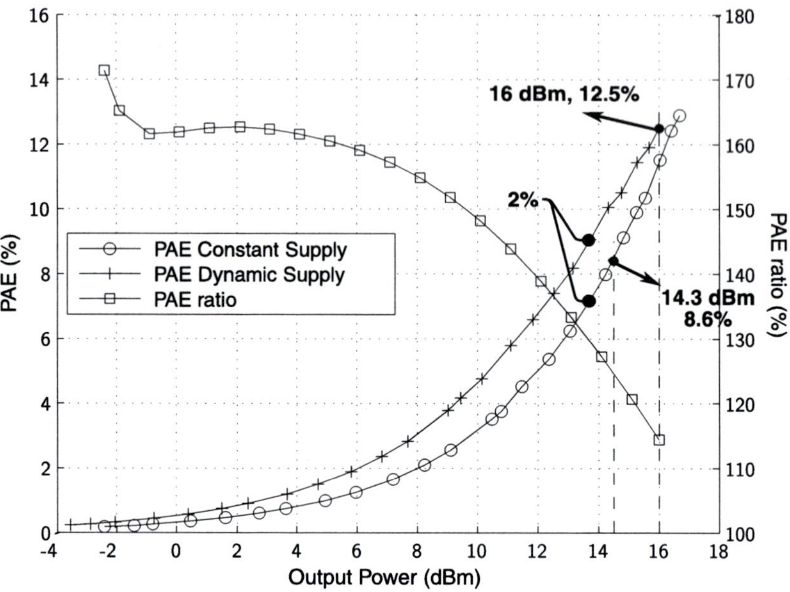

Fig. 4.16 Measured 2-tone PAE at 5.2 GHz

supply. However, when we consider the -35 dBc limit adopted (*dashed line*), the dynamic supply operation shows an extended linear range as already suggested by Fig. 4.14. The maximum linear output power is 16 dBm (40 mW) for the dynamic supply and 14.3 dBm (27 mW) for the constant supply.

The PAE[6] is shown in Fig. 4.16, revealing that for all the output power levels measured, the efficiency of the dynamic supply PA is higher than that of its constant supply counterpart. In order to have a quantitative indication of the improvement in efficiency, the ratio between the PAE with dynamic supply and constant supply is also shown in the figure (*right y-axis*, with 100% meaning equal efficiencies). For low output power levels, where the dynamic supply technique is more effective, a relative improvement of a factor of 1.7 (PAE ratio = 170%) is achieved. For an equal high linear output power, for example 13.7 dBm (23.5 mW), the absolute improvement in efficiency is 2%. At the -35 dBc linearity limit, the dynamic supply PA presents a PAE equal to 12.5%, whereas under constant supply operation it reaches 8.6%. It is worth emphasizing that the power consumption of the modulator is included in the PAE calculation. Its measured efficiency is between 65% at low input envelope amplitude levels and 86% at high levels.

---

[6]For the calculation of the PAE, the DC power consumption was obtained by multiplying the 2.5 V power supply voltage by the rms current delivered by it. The rms current was measured with a true-rms multimeter. The power delivered by the 2.5 V supply includes the DC consumption of both the modulator and the PA. The consumption of the envelope detection and processing blocks was not considered in the PAE calculation. For a detailed illustration of the measurement setup, refer to Chap. 9 (Appendix A).

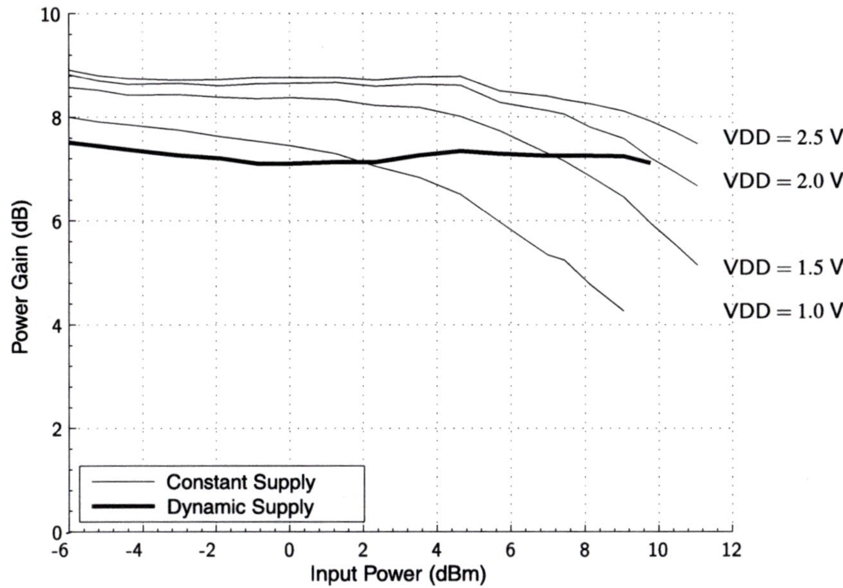

Fig. 4.17 Measured 2-tone power gain at 2.4 GHz

Measurements were also performed at 2.4 GHz. For this purpose, the evaluation board with the power amplifier matched at 2.4 GHz (refer to Sect. 4.4) was used. The power gain at different constant supply voltages—from 1 V to 2.5 V in 0.5 V steps— are shown in Fig. 4.17. The envelope processing circuit was calibrated based on these constant VDD curves. Figure 4.17 also shows the power gain of the PA under dynamic supply operation. Like in the 5.2 GHz case, this power gain is almost flat for varying input power indicating an extended linear operation with the dynamic power supply.

The IMD3 and PAE measurement results are shown in Figs. 4.18 and 4.19. Again, the dynamic supply operation generates higher distortion at low output power levels in contrast with the 2.5 V constant supply operation. However, when we consider the adopted −35 dBc IMD3 linearity limit, the dynamic supply operation provides and extended linear range, allowing the PA to reach 16.9 dBm (49 mW) against 15.9 dBm (38.9 mW) under constant supply operation.

The improvement in efficiency observed at 5.2 GHz is even more pronounced at 2.4 GHz. For low output power levels, a relative improvement of a factor of up to 2.1 (PAE ratio = 210%) is achieved. At an equal high output power level, for example 15.4 dBm (34.7 mW), an absolute improvement of 5.4% is observed with dynamic supply operation. At the −35 dBc linearity limit, the dynamic supply PA attains 22.5% against 15% of its constant supply counterpart.

For an easier comparison, the output power, PAE, and power gain values measured at the −35 dBc linearity limit for 2.4 and 5.2 GHz are presented in Table 4.4.

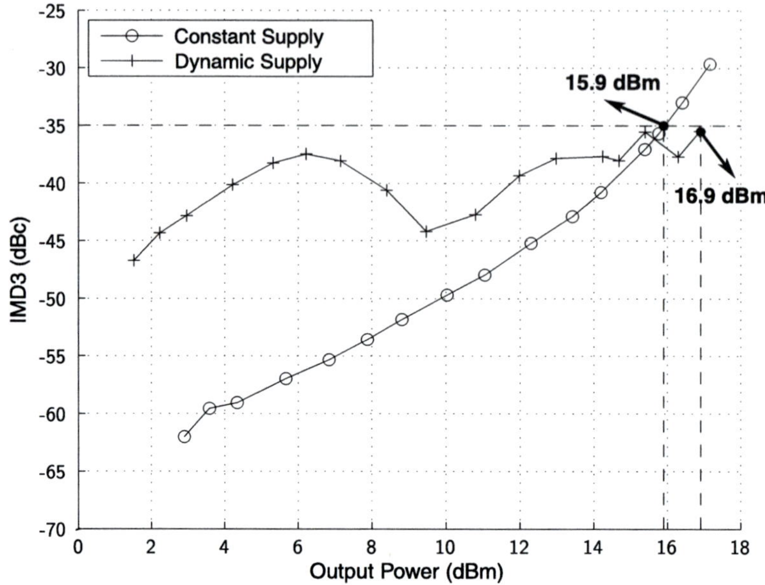

Fig. 4.18 Measured 2-tone IMD3 at 2.4 GHz. IMD3 limit of −35 dBc is also shown (*dashed line*)

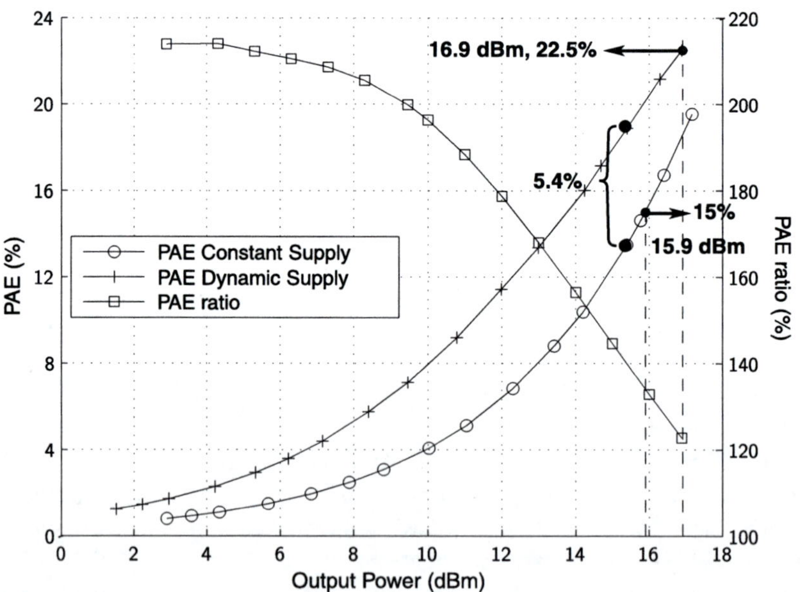

Fig. 4.19 Measured 2-tone PAE at 2.4 GHz

Table 4.4 Two-tone measurement results at −35 dBc IMD3

2 Tone	2.4 GHz		5.2 GHz	
	Const.	Dyn.	Const.	Dyn.
Output power (dBm)	15.9	16.9	14.3	16
PAE (%)	15	22.5	8.6	12.5
Gain (dB)	8.3	7.1	4.8	3.6

Table 4.5 Filter used for the OFDM measurements

Component	Value
L_{filter}	3.3 µH [6]
C_{filter}	27 pF [19]
R_{comp}	200 Ω
C_{comp}	4.7 nF
Cutoff frequency	3.8 MHz
f_s	42 MHz

4.5.2 OFDM Measurements

Due to the rapid envelope variation of an OFDM signal[7], a modification in the LC filter, according to the values shown in Table 4.5, was necessary to increase its cut-off frequency. A series RC compensation network (R_{comp}, C_{comp}) in parallel with the PA [28, Fig. 5.16] was used to modify Z_{PA}. Simulation results showed that a cutoff frequency of 3.8 MHz and a switching frequency of 42 MHz could be obtained.

The measurement results[8] presented in this section are only for 2.4 GHz. At 5.2 GHz, the envelope detector did not work properly with an OFDM signal, introducing too much delay in the envelope signal. Although at 2.4 GHz the envelope detector also introduced a delay, the delay was smaller and could be compensated with a delay line added in the RF signal path before the PA input. Screenshots of the oscilloscope showing this delay are provided in Chap. 9 (Appendix A).

Figure 4.20 shows a linearity measurement in terms of the rms value of the EVM. The input is an OFDM signal with a 64 Quadrature Amplitude Modulation (QAM) with a data rate of 54 Mbps, 3/4 coding rate, and 52 subcarriers. The IEEE 802.11a/g standards [12, 13] define a limit of 5.6% (−25 dB) for the rms value of the EVM. However, it is common to adopt a lower limit of 3% for the power amplifier to allow for the nonlinearity of other blocks in the transmitter. Figure 4.20 shows that the dynamic supply PA generates more distortion than its constant supply counterpart. At 3% EVM (*dashed line*), the dynamic supply attains an output

---

[7] The envelope bandwidth of an OFDM signal is 20 MHz.

[8] The setup used in the OFDM measurements is described in Chap. 9 (Appendix A).

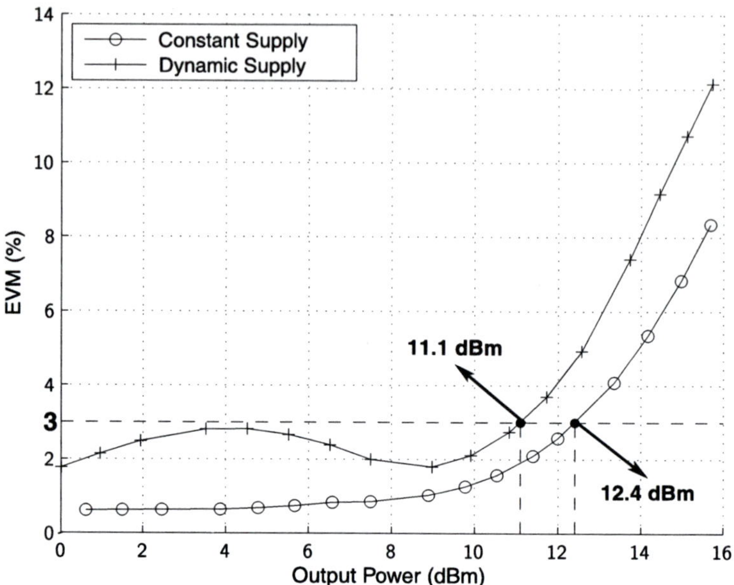

Fig. 4.20 Measured EVM (OFDM, 2.4 GHz). EVM limit of 3% is also shown (*dashed line*)

power of 11.1 dBm (12.9 mW), which is lower than that of the constant supply PA—
12.4 dBm (17.4 mW). Both output power levels are lower than the target 16 dBm
(40 mW).

Figure 4.21 shows the efficiency performance for the OFDM measurement. It
reveals that the dynamic supply PA presents a higher efficiency for all the output
power levels measured. At low output power levels, a relative improvement in efficiency of a factor of 2.4 (PAE ratio of 240%) is achieved. At an equal high linear
output power (less than 3% EVM), 11 dBm (12.6 mW) for example, the absolute
efficiency improvement is 3.2%. Figure 4.22 presents a plot comparing the PAE of
the dynamic and constant supply PAs in terms of EVM. It clearly shows that at 3%
EVM the dynamic supply RF PA presents a higher efficiency.

Table 4.6 summarizes the OFDM measurement results at 2.4 GHz. The target linear output power of 16 dBm (40 mW) could not be attained by neither the dynamic
supply nor the constant supply PA. We can conclude from this that the −35 dBc
IMD3 limit adopted in the design is not enough as a linearity requirement for WLAN
application. It is worth to note, however, that the envelope detection and processing blocks were implemented with discrete off-the-shelf components and, hence,
they contributed with some level of distortion. This explains the lower linear output
power level reached with the dynamic supply. In its application within a modern
transceiver, the envelope information would be readily available in the baseband
circuit and the envelope processing could be implemented much more effectively in
a DSP. This would isolate a possible linearity problem to be an issue related only
to the modulator and the PA. However, in our test environment, each of the blocks
used in the envelope path contributed with some distortion level.

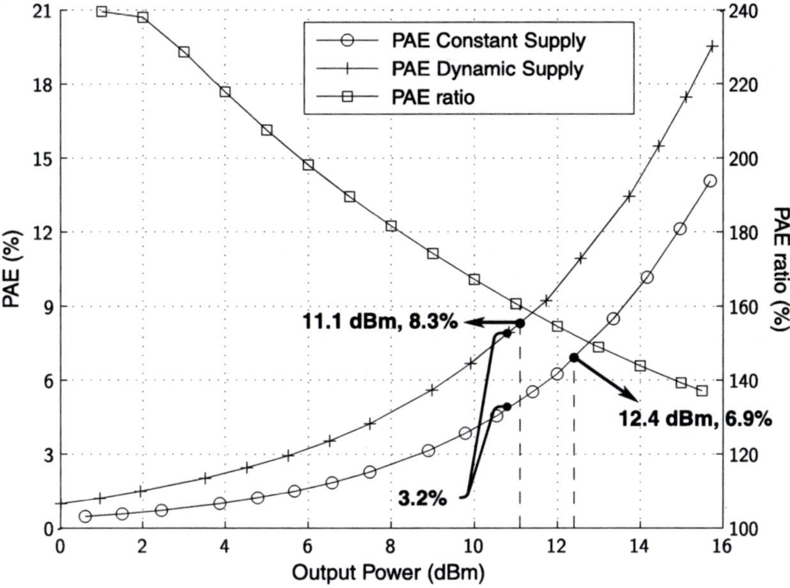

Fig. 4.21 Measured PAE (OFDM, 2.4 GHz)

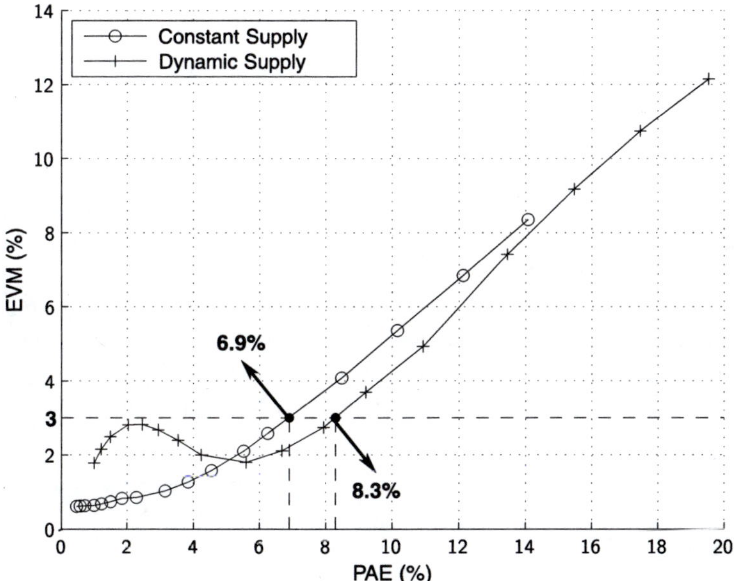

Fig. 4.22 Efficiency–linearity comparison (OFDM, 2.4 GHz). EVM limit of 3% is also shown (*dashed line*)

Table 4.6 OFDM measurement results at 2.4 GHz and 3% EVM

OFDM	Constant	Dynamic
Output power (dBm)	12.4	11.1
PAE (%)	6.9	8.3
Gain (dB)	7.9	8.6

Table 4.7 Comparison of the results with other works on dynamic supply PAs

Reference	[4]	[9]	[29]	[30]	[31]	This work	This work
Technique	Dyn.	Dyn.	Dyn.	Dyn.	Switched Dyn.	Dyn.	Dyn.
Technol. PA	Discr.	GaAs	Discr. Bip.	Discr. GaAs	Discr. Bip.	CMOS	CMOS
Technol. Mod.	Discr. GaAs	GaAs	CMOS	Discr. MOS	Discr. MOS	CMOS	CMOS
Supply (V)	–	3.6	3.3	6	5.5	2.5	2.5
f_c (GHz)	4	0.95	1.9	2.4	2.4	5.2	2.4
P_{out} (dBm)	36	20	20.5	15.1	19	16	16.9
PAE (%)	11.7	14	32	25	28	12.5	22.5
ACPR (dBc)	–	−26	−45	–	–	–	–
EVM (%)	–	–	4.9	3.2	2.8	–	–
IMD (dBc)	−50	–	−30	–	–	35	−35
Mod. Eff. (%)	–	74	85	75	60	86	86
f_s (MHz)	–	10	16	7	6	24	24
Linearization	–	–	–	DPD	DPD	–	–
Application	TCM	CDMA	CDMA	WLAN	WLAN	–	–

Note: Technol. = Technology; Mod. = Modulator; Eff. = Efficiency; Discr. = Discrete; Bip. = Bipolar

We can say that there are two main causes for the degradation in linearity with an OFDM signal input: the difference in the phase between the signals at the supply and at the output of the PA introduced mainly by the envelope detector, and the imperfect reproduction of the envelope signal at the PA supply due to the LC filter (low cutoff and switching frequencies). Nevertheless, Figs. 4.21 and 4.22 show that the dynamic supply technique provides a relatively large improvement in efficiency also for OFDM signals. This indicates that the dynamic supply PA can be considered for use in WLAN applications.

4.6 Conclusion

This chapter presented the measurement results obtained in the characterization of the dynamic supply RF power amplifier. The high-efficiency, fast modulator and the PA, which were integrated on the same chip, were characterized separately at a first time, and together afterwards. The complete system characterization still required the envelope detector and processing blocks that were also treated in this chapter. Two-tone measurement results of the whole system indicated that the dynamic supply PA can be employed in CMOS designs at high frequencies (2.4 and 5.2 GHz) to mitigate the inherent linearity–efficiency trade-off. The OFDM results revealed that the efficiency improvement achieved with the dynamic supply technique is promising for future developments for WLAN transceivers. For such applications, improvements in the PA design as well as in the circuits in the envelope path are still required.

Table 4.7 summarizes the results obtained with the dynamic supply CMOS RF PA presented in this chapter, which compares favorably with the results of other works on dynamic supply PAs. It shows that this is the first full-CMOS implementation published in the literature.

References

1. Agilent (1999) 8719D Network Analyzers. Agilent Technologies, USA. URL http://cp.literature.agilent.com/litweb/pdf/08720-90288.pdf
2. Agilent (2000) 8560 E-Series and EC-Series Spectrum Analyzers. Agilent Technologies, USA. URL http://cp.literature.agilent.com/litweb/pdf/08560-90158.pdf
3. Agilent (2007) 85052D 3.5mm economy calibration kit. User's and Service Guide. URL http://cp.literature.agilent.com/litweb/pdf/85052-90079.pdf
4. Buoli C, Abbiati A, Riccardi D (1995) Microwave power amplifier with 'envelope controlled' drain power supply. In: Eur Microw Conf (EuMC'95), Bologna, Italy, vol 1, pp 31–35
5. Burr-Brown (2004) OPA690—Wideband, voltage-feedback operational amplifier with disable. Data Sheet. URL http://focus.ti.com/lit/ds/symlink/opa690.pdf
6. Coilcraft (2005) Power chip inductors—1812PS series. Data Sheet. URL http://www.coilcraft.com/pdfs/1812ps.pdf
7. Dal Fabbro PA, Meinen C, Kayal M, Kobayashi K, Watanabe Y (2006) A dynamic supply CMOS RF power amplifier for 2.4 GHz and 5.2 GHz frequency bands. In: IEEE Radio Freq Integr Circuits Symp (RFIC'06), San Francisco, CA, pp 169–172
8. Eo Y, Lee K (2004) High efficiency 5 GHz CMOS power amplifier with adaptive bias control circuit. In: IEEE Radio Freq Integr Circuits Symp (RFIC'04), Fort Worth, TX, pp 575–578
9. Hanington G, Pin-Fan C, Asbeck PM, Larson LE (1999) High-efficiency power amplifier using dynamic power-supply voltage for CDMA applications. IEEE Trans Microw Theory Tech 47(8):1471–1476
10. Huber+Suhner (2007) 23 SMA-50-0-2/111 NE—Coaxial panel connector. Data Sheet. URL http://www.hubersuhner.com
11. Huber+Suhner (2007b) 82 SMB-50-0-1/111 NH—Straight PCB jack. Data Sheet. URL http://www.hubersuhner.com
12. IEEE Standard 802.11a (1999) IEEE standard for information technology–telecommunications and information exchange between systems–local and metropolitan area networks—specific requirements, Part 11: Wireless LAN medium access control (MAC) and physical layer (PHY) specifications - high-speed physical layer in the 5 GHz band
13. IEEE Standard 802.11g (2003) IEEE standard for information technology–telecommunications and information exchange between systems–local and metropolitan area networks—specific requirements, Part 11: Wireless LAN medium access control (MAC) and physical layer (PHY) specifications—amendment 4: Further higher data rate extension in the 2.4 GHz band

14. Kayal M (2002) Electronique I & II. Course Notes, Lausanne, Switzerland
15. Lee TH (1998) The Design of CMOS Radio-Frequency Integrated Circuits. Cambridge University Press, Cambridge
16. Linear Technology (2002) DC539A demo board quick start guide. Reference Design. URL http://www.linear.com
17. Linear Technology (2002) LTC5508—300 MHz to 7 GHz RF power detector with buffered output in SC70 package. Data Sheet. URL http://www.linear.com
18. Murata (2007) Murata Chip S-parameter and Impedance Library Version 3.12.0. URL http://www.murata.com/designlib/mcsil/index.html
19. Murata (2008) GRM1555C1H270JZ01—Monolithic ceramic capacitors (0402, C0G, 27pF, 50Vdc). Data Sheet. URL http://www.murata.com
20. Murata (2008) LQP15MN10NG02—Chip coils for high frequency film type—LQP15M series (0402 size). Data Sheet. URL http://www.murata.com
21. Murata (2008) LQP15MN6N8B02—Chip coils for high frequency film type—LQP15M series (0402 size). Data Sheet. URL http://www.murata.com
22. National (2005) LM134/LM234/LM334—3-terminal adjustable current sources. Data Sheet. URL http://cache.national.com/ds/LM/LM134.pdf
23. Philips (2005) 1PSxSB17—4 V, 30 mA low C_d Schottky barrier diode. Data Sheet. URL http://www.nxp.com
24. Pozar DM (1998) Microwave Engineering, 2nd edn. Wiley, New York
25. Rogers (2006) RO4000 Series High Frequency Circuit Materials. Data Sheet 92-004. URL http://www.rogerscorp.com/acm/literature.aspx
26. Rohde & Schwarz (2002) FSIQ7 Signal Analyzer. Rohde & Schwarz, Germany. URL http://www.rohde-schwarz.com
27. Rohde & Schwarz (2002) SMIQ06B Vector Signal Generator. Rohde & Schwarz, Germany. URL http://www.rohde-schwarz.com
28. Schlumpf N (2004) Adaptation dynamique de la compression d'un amplificateur RF pour des signaux modulés en amplitude et en phase. PhD thesis, EPFL, Lausanne, Switzerland. URL http://library.epfl.ch/theses/?nr=3020
29. Schlumpf N, Declercq M, Dehollain C (2004) A fast modulator for dynamic supply linear RF power amplifier. IEEE J Solid-State Circ 39(7):1015–1025
30. Wang F, Ojo A, Kimball D, Asbeck P, Larson L (2004) Envelope tracking power amplifier with pre-distortion linearization for WLAN 802.11g. In: IEEE MTT-S Int Microw Symp Dig (IMS'04), Fort Worth, TX, vol 3, pp 1543–1546
31. Wang F, Kimball DF, Popp JD, Yang AH, Lie DY, Asbeck PM, Larson LE (2006) An improved power-added efficiency 19-dBm hybrid envelope elimination and restoration power amplifier for 802.11g WLAN applications. IEEE Trans Microw Theory Tech 54(12):4086–4099

Chapter 5
Frequency-Tunable Capability

Abstract This chapter introduces the concept of frequency-tunable capability applied to RF power amplifiers, making some definitions and establishing the metrics for the evaluation of the design, which is the subject of Chap. 6. It also surveys the main techniques found in the literature that could be used in the design of the frequency-tunable amplifier. The advantages and drawbacks of each technique are discussed and the choice of the coupled-inductors technique is explained, together with the description of its use to implement a novel tunable output impedance matching network to be employed in multiband RF power amplifiers.

5.1 Introduction

The quest for tunable amplifiers is an old issue as stated by Castro in 1966 [10]:

> The frequency changes imposed by HF propagation conditions usually require retuning the transmitter, readjusting levels, changing of antennas, etc. The elimination of the human factor in performing these operations contributes to the avoiding of errors and presents operational economic advantages.

The previous chapters presented the design, implementation, and characterization of a dynamic supply CMOS RF power amplifier. Measurement results were shown for 2.4 and 5.2 GHz. To characterize the dynamic supply PA at these two frequencies, different impedance matching networks were implemented. *What if*[1] we could have a power amplifier operating optimally at both frequencies without the need of modifying its impedance matching network? One possible answer to this question is the frequency-tunable RF PA that we present hereafter.

---

[1] These two words, very often employed by Barrie Gilbert to explain his inventiveness, which drove him to very important contributions to the field of analog circuit design, are, according to him [19], *the most potent path to invention.* It also led us to this invention [30].

P.A. Dal Fabbro, M. Kayal, *Linear CMOS RF Power Amplifiers for Wireless Applications,* 61
Analog Circuits and Signal Processing,
DOI 10.1007/978-90-481-9361-5_5, © Springer Science+Business Media B.V. 2010

5.1.1 Scenario

Several wireless communication standards are available today. The frequency bands used are subject to different regulations. The consumer industry is demanding multi-standard, multifrequency RF PAs. Such devices are realized with either a wideband PA covering all the frequency bands of interest or with a narrowband PA whose center frequency can be adjusted while switching to a different standard. The later option is the underlying principle of the frequency-tunable RF PA.

The scenario can be visualized by imagining a person traveling with his/her Personal Digital Assistant (PDA) from Country A to Country B. When in the airport in Country A, the PDA must comply with Standard A so that it can have access to Internet. This standard has a set of requirements, one of them being Frequency A. After a long flight to Country B, the person would like to access his/her e-mails already in the airport. However, there, the wireless communication is regulated by Standard B, which operates at Frequency B. In order that the PDA can connect to Internet in these two situations, it must be equipped with a PA that can work satisfactorily at Frequency A and Frequency B. This is where a multifrequency PA plays its major role. However, in the case where such a device is not available the PDA must count on two PAs, one operating at Frequency A and the second at Frequency B. It is obvious that if a multifrequency PA exists, it is preferred because of the silicon area saving with a single PA and the simplicity of the system that does not need to switch from one PA to another.

5.2 Definitions

It is worth clarifying the meaning of the word *tunable* that is used hereafter. We mean by frequency-tunable power amplifier a PA whose main characteristics are adaptive with frequency. Ideally, the objective is that the power amplifier operates optimally in each of the frequency bands for which it is designed. We also employ the word *tunable* to qualify the element that enables this feature: the *tunable* inductor in the output impedance matching network. In this case it means that its inductance can be varied with frequency.

5.2.1 Tuning Range

To compare the performances of different tuning techniques, we define in this section the Tuning Range (TR) as a figure of merit. In the context of the frequency-tunable amplifier that we are presenting, the TR can be related to the center frequency (f_c), the transformation factor (m), and the inductance (L). In the literature, different definitions are used although no explicit formula is given. In this book, we define the TR as the relative variation of a parameter with respect to its center value:

$$TR = 100 \times \frac{\Delta x}{x_c}; \tag{5.1}$$

where x is the tunable parameter, Δx is the difference between the highest (x_{hi}) and the lowest (x_{lo}) value that x can take, and x_{c} is the center value between them. Hence, the tuning range can be rewritten as

$$TR = 100 \times \left(2\frac{x_{\text{hi}} - x_{\text{lo}}}{x_{\text{hi}} + x_{\text{lo}}} \right). \tag{5.2}$$

For instance, a PA whose center frequency (f_{c}) can be tuned from 2 to 2.5 GHz has a TR of 22.2%.[2]

5.3 Overview of the Existing Frequency-Tunable Techniques

This section covers the most important works found in the literature that represent the state of the art with respect to the techniques that can be applied in the development of a frequency-tunable power amplifier. We classify these works into two different groups: broadband and narrowband techniques. Broadband techniques have the objective of enlarging the bandwidth of an amplifier without changing its center frequency. They are presented in Sect. 5.3.1. Narrowband techniques have the goal of adjusting the center frequency of the amplifier while keeping a narrow bandwidth around this frequency. They are described in Sect. 5.3.2. The technique used in the frequency-tunable RF power amplifier described in the following chapters falls into the second group and, hence, is also introduced in Sect. 5.3.2.

5.3.1 Broadband Techniques

5.3.1.1 Lossy Matching

The use of lossy matching networks is a common technique employed in MESFET amplifiers in order to achieve a flat gain and a low Voltage Standing Wave Ratio (VSWR). By using a resistive matching network, the gain roll-off of the amplifier is compensated at the expense of a lower power gain and, hence, a lower efficiency.

Arell and Hongsmatip in [3] and Zhu et al. in [64] designed GaAs RF power amplifiers with an operating frequency from 2 to 6 GHz. The output power is in the order of 40 dBm (10 W) with a PAE > 20%, and a VSWR < 2.

Bahl in [4] presents a MESFET PA operating from 5 to 8.5 GHz and providing more than 33 dBm (2 W) output power, 15 dB gain, 31% PAE, and a VSWR better than 2.

The design equations for the implementation of lossy matching networks can be found in [25].

---

[2] $f_{\text{c\_c}} = 2.25$ GHz, $f_{\text{c\_lo}} = 2$ GHz, $f_{\text{c\_hi}} = 2.5$ GHz, and $\Delta f_{\text{c}} = 0.5$ GHz.

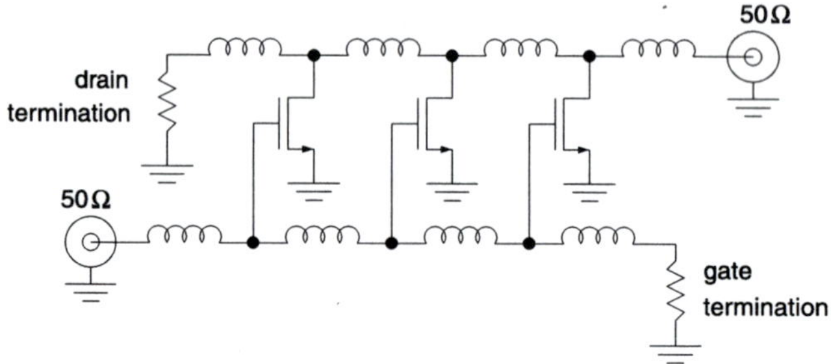

Fig. 5.1 Distributed amplifier configuration

5.3.1.2 Distributed Amplifiers

Another technique for broadband amplifier design is the distributed amplifier [42, Chap. 11], whose name was coined by Ginzton et al. in [20]. The structure of such amplifier is depicted in Fig. 5.1. Series inductors are used to separate the parasitic capacitance contribution of each gain stage, whereas the output currents are added at the output node. The gate inductors and the parasitic gate capacitors form an artificial gate transmission line, whereas the drain inductors and the parasitic drain capacitors form an artificial drain transmission line. The value of the characteristic impedance of these lines can be adjusted according to the terminal impedance to achieve a good matching across a broad frequency band. Distributed amplifiers were common in the design of GaAs amplifiers [37, 52] taking advantage of the high quality factor (Q) of the inductors that can be realized in these technologies.

A few designs of distributed amplifiers in CMOS technology were reported in the literature [5, 29, 33, 53], but the low quality factor of integrated inductors in CMOS processes is always a limiting factor. Sullivan et al. in [53] used package parasitics and bondwire inductances in order to circumvent this problem.

5.3.1.3 Balanced Amplifiers

The balanced amplifier [42, Chap. 11], [12, Chap. 13], [21, Chap. 4], [46, Chap. 15] shown in Fig. 5.2 is another broadband technique. A perfect matching is achieved at both the input and the output through the use of quadrature couplers. The two amplifiers in Fig. 5.2 are considered identical and their inputs are fed with signals that have half the power of the input signal and are 90° phase shifted from each other. Any reflection from the amplifiers input returns through the couplers and cancels at the RF input thanks to the phasing properties of the coupler. At the output, the signals are combined and the same reflection canceling mechanism operates. The bandwidth is limited by the quadrature couplers and, because each amplifier handles half of the power, the gain compression is improved by 3 dB. This kind of amplifier was first used in wideband microwave applications by Eisele et al. in [15], where a printed-circuit implementation reached 20 dB gain and 1.1 VSWR

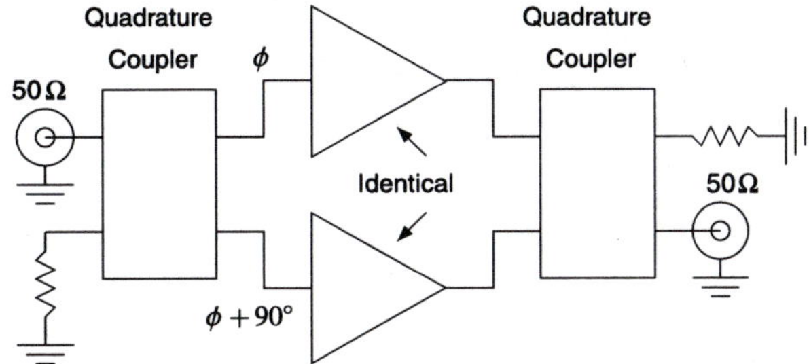

Fig. 5.2 Balanced amplifier block diagram

over a 600 MHz bandwidth. GaAs and CMOS implementations have been recently reported in [11] and [26], respectively. The main problem of balanced amplifiers is the need of quarter wavelength couplers that are usually not practical in integrated circuits at low gigahertz frequencies [61].

5.3.1.4 Drawbacks

The broadband techniques described in this section are not suitable for the design of the frequency-tunable CMOS RF PA. The lossy-matching technique decreases the power gain of the amplifier which is already low in CMOS designs. The distributed amplifier requires high-Q inductors difficult to realize with on-chip CMOS coils. Furthermore, the number of inductors required can be limiting in terms of silicon area. The balanced amplifier requires quadrature couplers which, besides having bandwidth limitations, require long transmission lines when operating at low giga-hertz frequencies, thereby occupying a large Si area.

5.3.2 Narrowband Techniques

In narrowband amplifiers, impedance matching is normally achieved with impedance matching networks composed of passive reactive elements. They achieve their best performance at the operating frequency for which they are designed. Outside the bandwidth for which the performance is optimum, the matching is poor and the amplifier is out of specifications.

In all the narrowband techniques covered in this chapter, at least one element of the matching network is tunable. The tunable element can be a capacitor or an inductor. In this section, we classify the narrowband techniques according to the way the value of this component is tuned:

- Varactors
  - Barium-Strontium-Titanate (BST) Capacitors [1, 38, 49, 55]
  - MOS Capacitors [43, 50]
  - Varactor Diodes [23, 36]

- Variable inductors
  - Ferromagnetic Inductors [56]
  - Micro-Electro-Mechanical System (MEMS) Inductors [34, 65]
  - Active Inductors [31, 35]
  - Saturable Reactors [45]
  - Coupled Inductors [6, 7, 13, 17, 18, 32, 41, 47, 51, 57]
- Switching
  - MEMS Switches [8, 40]
  - P-Intrinsic-N (PIN) Diodes [61, 62].

5.3.2.1 Varactors

The Britannica Encyclopedia online defines the term *varactor* as follows [9]:

> The varactor (variable reactor) is a device whose reactance can be varied in a controlled manner with a bias voltage. It is a p–n junction with a special impurity profile, and its capacitance variation is very sensitive to reverse-biased voltage. Varactors are widely used in parametric amplification, harmonic generation, mixing, detection, and voltage-variable tuning applications.

In the literature, the term *varactor* has not been exclusively used for varactor diodes, as defined by the Brittanica Encyclopedia, but also for other types of tunable capacitors whose reactance is controlled by a bias voltage. This is in agreement with the first phrase of the definition above and will be employed throughout this chapter. Therefore, we classified the tunable BST and MOS capacitors under varactors.

BST Capacitors   Tunable circuits can be built using a BST capacitor [1, 38]. The dielectric material used is a ferroelectric ceramic, $(BaSr)TiO3$, that exhibits an electric field-dependent dielectric constant (ε) [38]. Its capacitance can be adjusted with a DC voltage. Tunable matching networks using BST capacitors [55] and varactors [49] have been reported. Vicki Chen et al. in [55] presented a tunable matching network with transformation factors from 3.8 to 1.7 over a frequency range from 420 to 500 MHz. However, as this method relies on a specific technology not available in standard processes, it was not considered as an option for the design of the frequency-tunable CMOS amplifier described in this book.

MOS Capacitors   A MOS transistor can be used as a capacitor by connecting together its drain and source terminals and applying a voltage V_c between the gate terminal and these two terminals. If V_c is positive, the MOS capacitor works in inversion mode and if it is negative in accumulation mode. In between these two regions (refer to [60, Fig. 10.10]), close to the transistor threshold voltage (V_T), the capacitance undergoes a high variation with the applied voltage. By varying the applied voltage close to V_T the MOS capacitor can work as a varactor [60, Chap. 10].

In [50], a MOS varactor is used in the interstage matching of an integrated CMOS LNA tuning its operating frequency from 1.8 to 2.4 GHz (TR = 28.6%). In [43], a discrete MOSFET is used as a MOS capacitor in the design of a tunable class E PA operating in a frequency range of 19–31 MHz (TR = 48%).

Varactor Diodes   Varactor diodes are commonly used in Voltage-Controlled Oscillators (VCOs) [60, Chap. 2]. In the design of tunable matching networks, discrete implementations such as [23], and SiGe HBT implementations such as [36], have been reported. In [36], the power amplifier delivers 27 dBm (500 mW) output power in the 900, 1800, 1900, and 2100 MHz bands. In [23], the tunable matching network allows a complex impedance of magnitude from 6.3 to 1120 Ω to be matched at 1 GHz.

5.3.2.2 Variable Inductors

Several possibilities for inductor variation have been recently reported. These alternatives are analyzed in the following sections.

Ferromagnetic Inductors   FerroMagnetic (FM) materials can be used in inductor cores to form tunable inductors. By injecting a DC current in the inductor, the permeability of the core changes and so does the inductance. Vroubel et al. in [56], presented a tunable inductor comprising a planar solenoid with a thin film ferromagnetic NiFe core. With the inductance ranging from 1 to 150 nH, the authors demonstrated that a relative inductance variation ($\Delta L/L$) from 20% at 2 GHz up to 85% at 100 MHz could be achieved. The drawbacks are the low quality factor ($Q < 2$) and high DC current consumption.

MEMS Inductors   The advances in MEMS devices in the last decade have allowed their use in a number of applications. The field of tunable RF devices is continuously experiencing the benefits of such advances [16]. A tunable inductor can be built with MEMS technology as reported in [34, 65]. Zine-El-Abidine et al. in [65] proposed an inductor whose value was tunable by controlling the magnetic coupling between two inductors. This magnetic coupling contributes to the total inductance and is controlled by varying the distance between the two inductors using thermal MEMS actuators. An inductor with a TR of 12.5% in a frequency range of 2–5 GHz could be achieved. In [34], the angle between two inductors is controlled with some kind of self-assembling technique allowing 18% inductance variation (TR = 20%), over 15 GHz Self-Resonant Frequency (SRF) and a quality factor higher than 13. Okada et al. in [39] presented an inductor whose value could be tuned by the movement of a metal plate controlled by a MEMS actuator. However, the actuator itself was not implemented and the movement of the plate was made manually by a micromanipulator.

Active Inductors   Active inductors are another possible implementation of tunable inductors, but high consumption, complexity, noise, and nonlinearity [56] impeded their widespread. Active inductors are mainly designed based on the gyrator principle, like in [58], or Generalized Impedance Converter (GIC) topologies, as in [22, 31, 35, 59]. Kobayashi et al. in [31] proposed an active inductor using the common-gate, cascode-FET feedback topology [22] to implement a GaAs VCO-mixer with a tuning range of 28.5–29.3 GHz (TR $= 2.8\%$). Mukhopadhyay et al. in [35] developed a CMOS regulated cascode active inductor [54] for a VCO. The inductor could be varied from 0.1 to 15 nH at frequencies from 0.5 to 3 GHz with a quality factor higher than 35. A VCO implemented with this inductor attained a TR of 120%. Although the values reported for quality factor, inductance, and frequency ranges are very attractive, active inductors present limitations in terms of linearity and power handling capability which are very important in power amplifier applications.

Saturable Reactors A special type of tunable inductor called saturable reactor [44] was reported in [45]. In a similar way to the ferromagnetic inductor, the principle of the saturable reactor is based on the variation of the permeability of its core by the application of a DC bias field. Several possibilities of usage of this inductor together with variable capacitors are presented in [45] in order to implement a frequency-agile PA working in frequencies of 5 to 21 MHz.

Coupled Inductors The use of coupled inductors as a tunable inductor was first reported in 1949 by Johnson in [27, 28]. The tunable inductance was used in a pentode valve frequency-modulated oscillator whose resonance frequency could be tuned within a range of 30%. This kind of oscillator was used as a wobbulator[3] for receiver alignment. The integrated version of the coupled-inductors-based tunable inductance would be reported for the first time only almost 50 years later by Pehlke et al. in [41].

Figure 5.3 illustrates the concept of how an inductor can be tuned by exploiting the magnetic coupling between two identical inductors. We apply a control current (I_{ctrl}) through L_2 of the same amplitude of the RF current (I_{RF}) through L_1. If the phase shift between I_{ctrl} and I_{RF} is zero, the total inductance seen by the RF circuit connected to L_1 will be

$$L_{eq} = L_1 + M, \tag{5.3}$$

where M is the mutual inductance between L_1 and L_2 and is calculated as

$$M = k\sqrt{L_1 L_2}, \tag{5.4}$$

where k is the coupling factor between L_1 and L_2 and is considered equal to unity in this simple reasoning. On the other hand, if the phase shift between these two currents equals $180°$, the total inductance will be

$$L_{eq} = L_1 - M. \tag{5.5}$$

Hence, if the phase shift between these two currents is varied, it is possible to tune the total inductance seen by the RF circuit from $L_1 - M$ to $L_1 + M$.

---

[3] Wobbulator is a sweep signal generator that was used in cathode-ray oscilloscopes and as the local oscillator in panoramic superheterodyne receivers.

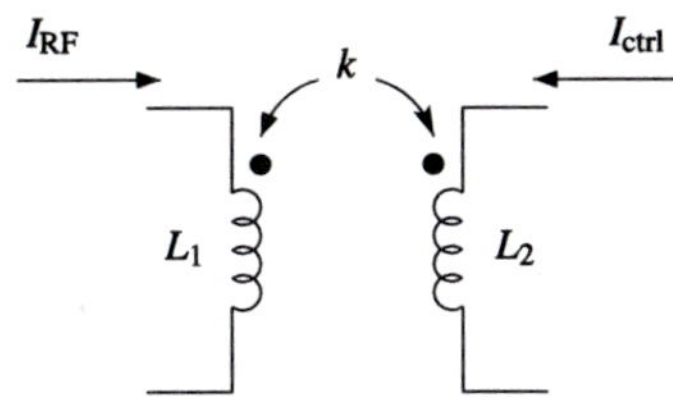

Fig. 5.3 Coupled inductors used as a tunable inductance

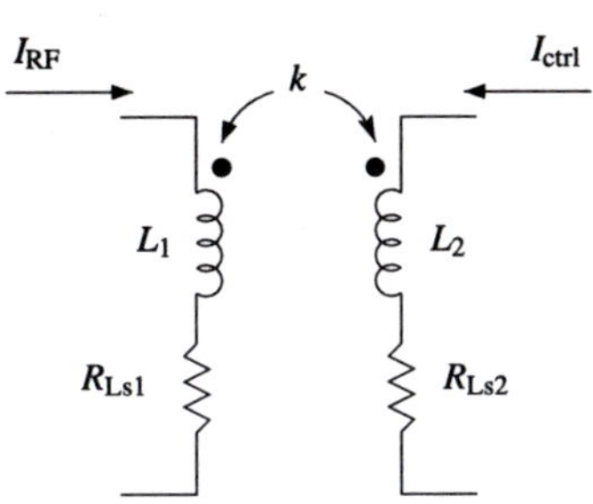

Fig. 5.4 Coupled inductors used as a tunable inductance (parasitic series resistances included)

To make the inductors of Fig. 5.3 more realistic, a parasitic series resistance is included as shown in Fig. 5.4. As a result, when we inject a current I_{ctrl} through L_2, besides the change in the inductance, a resistive part appears in series with the impedance seen by the RF circuit connected to L_1. By varying the amplitude and phase of I_{ctrl}, one can find adequate values for which the resistive part added by the mutual effect is negative and cancels (or at least decreases) the original parasitic series resistance (R_{Ls1}) of L_1.

The relationship between the currents through L_1 and L_2 can be expressed, in rectangular notation, by

$$\frac{I_{\text{ctrl}}}{I_{\text{RF}}} = \alpha + i\beta, \tag{5.6}$$

where α and β denote the real and imaginary parts of the current ratio. In polar notation, (5.6) becomes

$$\frac{I_{\text{ctrl}}}{I_{\text{RF}}} = r(\cos\phi + i\sin\phi), \tag{5.7}$$

where r is the magnitude and ϕ is the phase of the ratio between I_{ctrl} and I_{RF}. Alternatively, r can be seen as the attenuation in the magnitude of I_{ctrl} when compared to I_{RF} and ϕ as the phase shift between these two currents.

If we consider that the inductors in Fig. 5.4 have the same inductance $L_1 = L_2 = L$ and the same series parasitic resistance $R_{\text{Ls1}} = R_{\text{Ls2}} = R_{\text{Ls}}$, the impedance seen by the RF circuit is

$$Z_{\text{eq}} = -\omega k\beta L + R_{\text{Ls}} + i[\omega L(1+\alpha k)], \tag{5.8}$$

which can be split into a resistance and a reactance:

$$R_{\text{eq}} = -\omega k\beta L + R_{\text{Ls}}, \tag{5.9}$$

$$X_{\text{eq}} = \omega L(1+\alpha k). \tag{5.10}$$

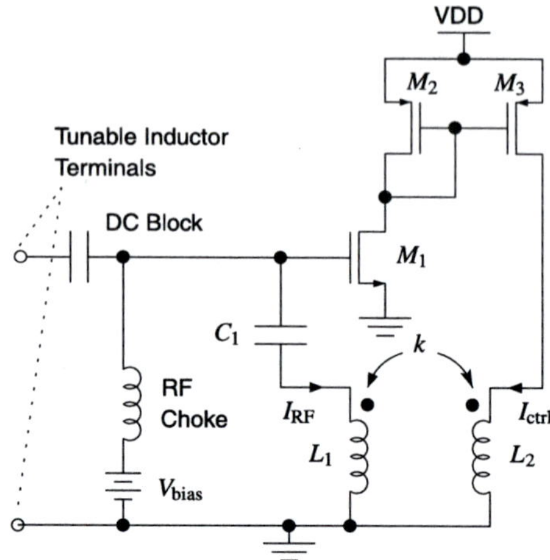

Fig. 5.5 Example of circuit for current control used in [57]

The equivalent inductance seen by the RF circuit and the corresponding quality factor are

$$L_{eq} = L(1 + \alpha k), \tag{5.11}$$

$$Q_{eq} = \frac{\omega L(1 + \alpha k)}{R_{Ls}(1 - \beta k Q_u)}, \tag{5.12}$$

where $Q_u = \omega L / R_{Ls}$ is the unloaded quality factor of inductors L_1 and L_2.

Equations (5.9), (5.11) and (5.12) can be written in polar notation as follows:

$$R_{eq} = -\omega L k r \sin\phi + R_{Ls}, \tag{5.13}$$

$$L_{eq} = L(1 + kr\cos\phi), \tag{5.14}$$

$$Q_{eq} = \frac{\omega L(1 + kr\cos\phi)}{R_{Ls}(1 - kr\sin\phi Q_u)}. \tag{5.15}$$

Equations (5.14) and (5.15) show that the equivalent inductance and quality factor are determined by the choice of the magnitude and the phase of the ratio between the currents flowing through the inductors—refer to (5.7). Pehlke et al. in [41] demonstrated, with a measurement setup including a directional coupler, an amplifier, a variable attenuator, and a variable phase shifter, that an inductor with a TR of 67% (0.8–1.6 nH) could be achieved with integrated silicon coupled inductors at 2 GHz. They also showed that a quality factor of 2000 could be reached at a particular amplitude and phase of the control current. The coupled inductors were integrated in a 0.18 μm CMOS technology and a photograph of the tunable inductor can be found in [41, Fig. 2].

Although in [41] tuning and Q enhancement were demonstrated, a current control circuit viable for an IC implementation was not presented. Wu and Chang in [57] proposes a CMOS control circuit (Fig. 5.5) to use the coupled-inductor technique exclusively as a Q-enhancement technique. The input voltage applied at the tunable inductor terminals appears equally at the gate of M_1 and at the primary winding of the coupled inductors—capacitor C_1 acts like a DC block—generating the current I_{RF}. The drain current of M_1 is related to the input voltage through the transconductance g_{m1}. This current generates a voltage at the gate of transistors M_2 and M_3 when flowing through $1/g_{m2}$. The current through M_3 is related to its gate voltage through g_{m3} and is the same current flowing into the secondary winding of the coupled inductors. Hence, the relationship between I_{ctrl} and I_{RF} depends on the transconductances g_{m1}, g_{m2}, g_{m3}, and on the inductance L_1. The correct adjustment of these parameters leads to the cancellation of the real part of the impedance seen by the RF circuit (refer to [57, (3)]). The circuit was implemented in a 0.18 μm CMOS technology and quality factor values higher than 3000 were reported from 1.5 to 2.1 GHz.

The coupled-inductor technique has been used in integrated tunable RF circuits since its first demonstration in 1997. In [6, 7], using three inductors with magnetic coupling between each two of them, the authors developed a tunable passband CMOS RF filter. The filter attained a blocking dynamic range of 80 dB with a center frequency with a TR of 11% around 1 GHz. Other applications using the coupled-inductor technique in Q-enhanced RF filters have been reported [17, 18, 32, 51]. The application in tunable oscillators was described in [13, 47]. It is worth noting that the coupled-inductor technique has also been subject of research under the name of Boot-Strapped Inductor (BSI), as found in [2, 14, 48].

5.3.2.3 Switching

Tunable matching networks can also have recourse to a switching strategy to vary the capacitance or inductance of one or more network branches. Two kinds of integrated switching strategies can be found in the literature and are explained below.

MEMS Switches Another use of MEMS in tunable matching networks, other than the inductors described in Sect. 5.3.2.2, is the MEMS switch. Bartlett et al. in [8] use these switches within a bank of parallel inductors in the series branch and a bank of parallel capacitors in the shunt branch of an L-type output matching network of a PA. Zhou et al. in [63] implemented variable inductors that could achieve a tuning range from 2.5 to 324.8 nH (TR = 197%) in frequencies from 0.5 to 1.6 GHz using four MEMS switches. In [40], the switches are used to activate stub tuners that can be used in impedance matching networks. The tuners can operate from 10 to 20 GHz to match loads with real parts between 1.5 and 110 Ω and imaginary parts between -260 and 91 Ω.

PIN Diodes PIN diodes have a structure in which a lightly doped (nearly intrinsic) semiconductor separates a p^+ from a n^+ region. The device behaves as a voltage controlled resistance that can be used as a switch in RF frequencies [24, p. 904]. The resistance of a forward-biased PIN diode is controlled by the DC current flowing through it. In [61, 62], a PIN diode, four capacitors, and four inductors form a π-matching network. The series inductance between the two shunt capacitors of the network is varied by putting inductors in parallel when the PIN diode turns on. The GaAs HBT PA is able to operate in the 900 MHz band when the diode is off and at 1.8 GHz when it is on. The output power at the 1 dB compression point is above 30 dBm (1 W) with a corresponding PAE of 40%.

5.4 Conclusion

This chapter provided an overview of the main existing tuning techniques that could be used in the development of a frequency-tunable CMOS RF power amplifier. They were divided in broadband and narrowband techniques and they can serve as a reference for researchers working with tunable circuits.

In Table 5.1, the advantages and drawbacks of each technique are displayed. The MOS capacitor and varactor diode present no major drawback. However, as it will be seen in the next chapter, a variable inductor is preferred to a variable capacitance as the tunable element. For the frequency-tunable CMOS RF PA herein described, the coupled inductors technique is the best candidate mainly because of the possible Q enhancement necessary to compensate for the low Q of integrated inductors in CMOS technology. This results in a power amplifier with improved efficiency. Furthermore, it presents no major constraints regarding silicon area, complexity, power handling, linearity, or tuning range. (The only drawback concerns stability, but potential stability problems can be detected and solved in the design phase.) No RF power amplifiers using this technique have been reported so far. In the next chapter, the design of a frequency-tunable CMOS RF PA based on coupled inductors will be described.

Table 5.1 Comparison among the tuning techniques

Technique	Advantages	Drawbacks
Lossy matching	Simplicity, bandwidth	Efficiency
Distributed amplifier	Bandwidth	High-Q inductor, area
Balanced amplifier	Bandwidth	Hybrid couplers
BST capacitors	Tuning range	Integration
FM inductors	Tuning range	Integration
MEMS inductors	Efficiency	Integration, tuning range
Active inductors	Area, tuning range	Complexity, power handling, efficiency
Saturable reactors	Tuning range	Integration
Coupled inductors	Efficiency, area, Q enhancement	Stability
PIN diodes	Power handling	Efficiency
MEMS switches	Efficiency	Integration

References

1. Agile Materials (2004) Tunability—An enabling technology for wireless. White Paper. URL http://www.agilerf.com/pdf/Tunability_WhitePaper.pdf
2. Albertoni F, Fanucci L, Neri B, Sentieri EA (2001) Tuned LNA for RFICs using boot-strapped inductor. In: IEEE Radio Freq Integr Circuits Symp (RFIC'01), Phoenix, AZ, pp 83–86
3. Arell T, Hongsmatip T (1993) A unique MMIC broadband power amplifier approach. IEEE J Solid-State Circ 28(10):1005–1010
4. Bahl IJ (2004) Low loss matching (LLM) design technique for power amplifiers. IEEE Microw Mag 5(4):66–71
5. Ballweber BM, Gupta R, Allstot DJ (2000) A fully integrated 0.5–5.5 GHz CMOS distributed amplifier. IEEE J Solid-State Circ 35(2):231–239
6. Bantas S, Koutsoyannopoulos Y (2004) CMOS active-LC bandpass filters with coupled-inductor Q-enhancement and center frequency tuning. IEEE Trans Circ Syst II 51(2):69–76
7. Bantas S, Papananos Y, Koutsoyannopoulos Y (1999) Cmos tunable bandpass RF filters utilizing coupled on-chip inductors. In: Proc Int Symp Circ and Syst (ISCAS'99), Orlando, FL, vol 2, pp 581–584
8. Bartlett JL, Chang MCF, Marcy HO, Pedrotti KD, Pehlke DR, Seabury CW, Yao JJ, Mehrotra D, Tham JLJ (2001) Integrated tunable high efficiency power amplifier. US Patent
9. Britannica (2010) Brittanica encyclopedia online. URL http://www.britannica.com/
10. Castro A (1966) Automatic tuning system for high-power amplifiers. IEEE Trans Commun Technol 14(6):824–834
11. Cegielski T, Matuszewski R (2004) The design of medium power C-band balanced amplifiers. In: Int Conf Microw Radar Wirel Commun (MIKON'04), Warsaw, Poland, vol 1, pp 107–110
12. Cripps SC (2006) RF Power Amplifiers for Wireless Communications, 2nd edn. Artech House, Norwood
13. Cusmai G, Repossi M, Albasini G, Svelto F (2007) A 3.2-to-7.3 GHz quadrature oscillator with magnetic tuning. In: IEEE Int Solid-State Circuits Conf Dig Tech Pap (ISSCC'07), San Francisco, CA, pp 92–93, 589
14. D'Angelo G, Fanucci L, Monorchio A, Monterastelli A, Neri B (1999) High-quality active inductors. Electron Lett 35(20):1727–1728
15. Eisele K, Engelbrecht R, Kurokawa K (1965) Balanced transistor amplifiers for precise wideband microwave applications. In: IEEE Int Solid-State Circuits Conf Dig Tech Pap (ISSCC'65), San Francisco, CA, vol VIII, pp 18–19
16. Fernández-Bolaños M, Lisec T, Dainesi P, Ionescu AM (2008) Thermally stable distributed MEMS phase shifter for airborne and space applications. In: Eur Microw Conf (EuMC'08), Amsterdam, The Netherlands, pp 100–103
17. Gee WA, Allen PE (2007) Cmos integrated LC RF bandpass filter with transformer-coupled Q-enhancement and optimized linearity. In: Proc Int Symp Circuits and Syst (ISCAS'07), New Orleans, LA, pp 1445–1448
18. Georgescu B, Pekau H, Haslett J, McRory J (2003) Tunable coupled inductor Q-enhancement for parallel resonant LC tanks. IEEE Trans Circ Syst II 50(10):750–713
19. Gilbert B (2003) The beginning of translinear circuit design. IEEE Solid-State Circ Soc Newsl 8(1):6–7
20. Ginzton EL, Hewlett WR, Jasberg JH, Noe JD (1948) Distributed amplification. Proc IRE 36(8):956–969
21. Gonzalez G (1997) Microwave Transistor Amplifiers: Analysis and Design, 2nd edn. Prentice-Hall, Upper Saddle River
22. Hara S, Tokumitsu T, Aikawa M (1989) Lossless broad-band monolithic microwave active inductors. IEEE Trans Microw Theory Tech 37(12):1979–1984
23. Hoarau C, Bailly PE, Arnould JD, Ferrari P, Xavier P (2007) A RF tunable impedance matching network with a complete design and measurement methodology. In: Eur Microw Conf (EuMC'07), Munich, Germany, pp 751–754

24. Horowitz P, Hill W (1989) The Art of Electronics, 2nd edn. Cambridge University Press, Cambridge
25. Ikalainen PK (1989) An RLC matching network and application in 1–20 GHz monolithic amplifier. In: 1989 IEEE MTT-S Int Microw Symp Dig (IMS'89), Long Beach, CA, vol 3, pp 1115–1118
26. Jin JD, Hsu SSH (2008) A 0.18-μm CMOS balanced amplifier for 24-GHz applications. IEEE J Solid-State Circ 43(2):440–445
27. Johnson KC (1949) Single-valve frequency-modulated oscillators—2.—Practical details of design and use. Wirel. World—Radio Electron LV(5):168–170
28. Johnson KC (1949) Single-valve frequency-modulated oscillators—new principle giving wide coverage. Wirel World—Radio Electron LV(4):122–123
29. Ker MD, Hsiao YW, Kuo BJ (2005) ESD protection design for 1 to 10-GHz distributed amplifier in CMOS technology. IEEE Trans Microw Theory Tech 53(9):2672–2681
30. Kobayashi K, Watanabe Y, Dal Fabbro P, Kayal M (2009) Tunable impedance matching circuit. International Patent, wO 2009/034659 A1
31. Kobayashi KW, Oki AK, Umemoto DK, Block TR, Streit DC (1998) A novel self-oscillating HEMT-HBT cascode VCO-mixer using an active tunable inductor. IEEE J Solid-State Circ 33(6):870–876
32. Kulyk J, Haslett J (2006) A monolithic CMOS 2368 $\pm$ 30 MHz transformer based Q-enhanced series-C coupled resonator bandpass filter. IEEE J Solid-State Circ 41(2):362–374
33. Liu RC, Lin CS, Deng KL, Wang H (2004) Design and analysis of DC–to–14-GHz and 22-GHz CMOS cascode distributed amplifiers. IEEE J Solid-State Circ 39(8):1370–1374
34. Lubecke VM, Barber B, Chan E, Lopez D, Gross ME, Gammel P (2001) Self-assembling MEMS variable and fixed RF inductors. IEEE Trans Microw Theory Tech 49(11):2093–2098
35. Mukhopadhyay R, Park Y, Lee CH, Nuttinck S, Laskar J (2004) Frequency-agile CMOS RFICs for multi-mode RF front-end. In: Proc Eur Conf Wirel Technol, Amsterdam, The Netherlands, pp 9–12
36. Neo WCE, Lin Y, Liu XD, De Vreede LCN, Larson LE, Spirito M, Pelk MJ, Buisman K, Akhnoukh A, De Graauw A, Nanver LK (2006) Adaptive multi-band multi-mode power amplifier using integrated varactor-based tunable matching networks. IEEE J Solid-State Circ 14(9):2166–2176
37. Niclas KB, Wilser WT, Kritzer TR, Pereira RR (1983) On theory and performance of solid-state microwave distributed amplifiers. IEEE Trans Microw Theory Tech 83(6):447–456
38. Noren B (2004) Thin film Barium Strontium Titanate (BST) for a new class of tunable RF components. Microw J 47:210–220
39. Okada K, Sugawara H, Ito H, Itoi K, Sato M, Abe H, Ito T, Masu K (2006) On-chip high-Q variable inductor using wafer-level chip-scale package technology. IEEE Trans Electron Devices 53(9):2401–2406
40. Papapolymerou J, Lange KL, Goldsmith CL, Malczewski A, Kleber J (2003) Reconfigurable double-stub tuners using MEMS switches for intelligent RF front-ends. IEEE Trans Microw Theory Tech 51(1):271–278
41. Pehlke DR, Burstein A, Chang MF (1997) Extremely high-Q tunable inductor for Si-based RF integrated circuit applications. In: IEEE Int Electron Devices Meet Tech Dig (IEDM'97), Washington, DC, pp 63–66
42. Pozar DM (1998) Microwave Engineering, 2nd edn. Wiley, New York
43. Raab FH (2001) Electronically tunable class-E power amplifier. In: 2001 IEEE MTT-S Int Microw Symp Dig (IMS'01), Phoenix, AZ, vol 3, pp 1513–1516
44. Raab FH (2007) Electronically tuned power amplifier. US Patent
45. Raab FH, Ruppe D (2003) Frequency-agile class-D power amplifier. In: Int Conf HF Radio Syst Tech, University of Bath, UK, pp 81–85
46. Radmanesh MM (2001) Radio Frequency and Microwave Electronics Illustrated. Prentice-Hall, Upper Saddle River
47. Rong S, Luong HC (2007) A 1 V 4 GHz-and-10 GHz transformer-based dual-band quadrature VCO in 0.18 μm CMOS. In: Proc IEEE Cust Integr Circuit Conf (CICC'07), San Jose, CA, pp 817–820

48. Scandurra G, Ciofi C, Zito D (2005) A new topology for transformer based CMOS active inductances. In: PhD Res Microelectron Electron (PRIME'05), Lausanne, Switzerland, vol 1, pp 27–30
49. Scheele P, Goelden F, Giere A, Mueller S, Jakoby R (2005) Continuously tunable impedance matching network using ferroelectric varactors. In: 2005 IEEE MTT-S Int Microw Symp Dig (IMS'05), Long Beach, CA, pp 603–606
50. Shin SH, Yoo HJ (2007) A multistandard RF front-end using varactor controlled tunable interstage matching network. In: IEEE Radio Wirel Symp (RWS'07), Long Beach, CA, pp 181–184
51. Soorapanth T, Wong SS (2002) A 0-dB IL 2140+30 MHz bandpass filter utilizing Q-enhanced spiral inductors in standard CMOS. IEEE J Solid-State Circ 37(5):579–586
52. Strid EW, Gleason KR (1982) A DC–12 GHz monolithic GaAsFET distributed amplifier. IEEE Trans Microw Theory Tech MTT-30(7):969–975
53. Sullivan PJ, Xavier BA, Ku WH (1997) An integrated CMOS distributed amplifier utilizing packaging inductance. IEEE Trans Microw Theory Tech 45(10):1969–1976
54. Thanachayanont A, Payne A (1996) VHF CMOS integrated active inductor. Electron Lett 32(11):999–1000
55. Vicki Chen LY, Forse R, Chase D, York RA (2004) Analog tunable matching network using integrated thin-film BST capacitors. In: 2004 IEEE MTT-S Int Microw Symp Dig (IMS'04), Fort Worth, TX, vol 1, pp 261–264
56. Vroubel M, Yan Z, Rejaei B, Burghartz JN (2004) Integrated tunable magnetic RF inductor. IEEE Electron Device Lett 25(12):787–789
57. Wu YC, Chang MF (2002) On-chip high-Q ($>$3000) transformer-type spiral inductors. Electron Lett 38(3):112–113
58. Xiao H, Schaumann R, Daasch WR, Wong PK, Pejcinovic B (2004) A radio-frequency CMOS active inductor and its application in designing high-Q filters. In: Proc Int Symp Circuits and Syst (ISCAS'04), Vancouver, Canada, vol 4, pp 197–200
59. Yodprasit U, Ngarmnil J (2000) Q-enhancing technique for RF CMOS active inductor. In: Proc Int Symp Circuits and Syst (ISCAS'00), Geneva, Switzerland, vol 5, pp 589–592
60. Ytterdal T, Cheng Y, Fjeldly TA (2003) Device Modeling for Analog and RF CMOS Circuit Design. Wiley, Chichester
61. Zhang H, Gao H, Li GP (2005) Broad-band power amplifier with a novel tunable output matching network. IEEE Trans Microw Theory Tech 53(11):3606–3614
62. Zhang H, Gao H, Li GP (2005) A novel tunable broadband power amplifier module operating from 0.8 GHz to 2.0 GHz. In: 2005 IEEE MTT-S Int Microw Symp Dig (IMS'05), Long Beach, CA, pp 661–664
63. Zhou S, Sun XQ, Carr WN (1997) A micro variable inductor chip using MEMS relays. In: Int Conf Solid-State Sens Actuators (TRANSDUCERS'97), Chicago, IL, vol 2, pp 1137–1140
64. Zhu X, Chen X, Ling J (2000) 2–6 GHz GaAs MMIC power amplifier. In: Int Conf Microw Millim Wave Technol (ICMMT'00), Beijing, China, pp 134–137
65. Zine-El-Abidine I, Okoniewski M, McRory JG (2003) A new class of tunable RF MEMS inductors. In: Int Conf MEMS NANO Smart Syst (ICMENS'03), Banff, Canada, pp 114–115

Chapter 6
Design of the Frequency-Tunable CMOS RF Power Amplifier

Abstract This chapter presents the design of an integrated frequency-tunable RF power amplifier for operation in the 3.7 and 5.2 GHz frequency bands. A novel tunable impedance matching network based on coupled inductors is employed at the output of the amplifier. The design of the complete system is described, which includes the integrated planar coupled inductors, the circuit that controls the current flowing through the inductors, and the RF power amplifier. Simulation results for the complete system integrated in a CMOS technology are presented.

6.1 Introduction

It was shown in the previous chapters that the output impedance matching network is of fundamental importance for the performance of the PA and that the variation in frequency alters the properties of this network, thereby changing the performance of the PA. Hence, a careful analysis of the output impedance matching is carried out in the Sect. 6.2 while the input matching network is subject of Sect. 6.3. The design of the RF PA, the coupled inductors, and the control circuit required to implement the tunable matching network based on coupled inductors are the subject of Sect. 6.4.

6.2 Output Impedance Matching

The function of the output matching network is to transform the antenna impedance into the desired impedance seen by the output stage of the PA. The illustration of a basic CMOS power amplifier circuit presented in Fig. 3.11 on p. 26 is repeated in Fig. 6.1 for clarity. This desired impedance can be the complex conjugate of the output impedance of the PA. This is the case of the *conjugate match*. On the other hand, the desired impedance can be the optimum resistance (R_{opt}). This is the case of the *power match* [6], which leads to an optimum performance at maximum output power. This performance is a trade-off between efficiency and linearity.

P.A. Dal Fabbro, M. Kayal, *Linear CMOS RF Power Amplifiers for Wireless Applications,*   77
Analog Circuits and Signal Processing,
DOI 10.1007/978-90-481-9361-5_6, © Springer Science+Business Media B.V. 2010

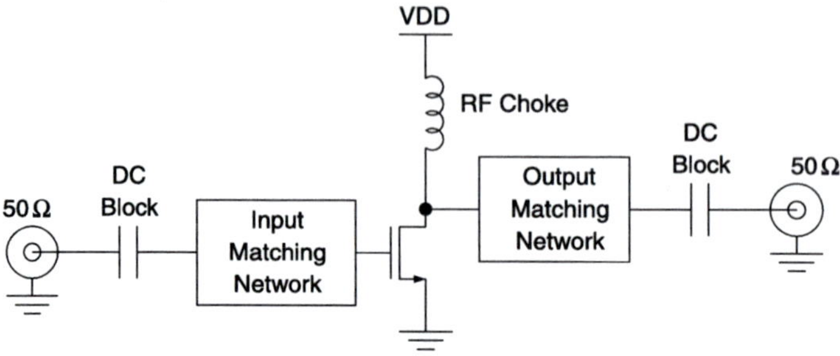

Fig. 6.1 Basic power amplifier structure

In RF PAs, a power match is generally sought at the output. In this case, the resulting R_{opt} takes into account the limitations of the output transistor with respect to current capability and maximum voltage swing. Another non-ideality of the transistor is its knee voltage (V_{knee}) [6]. It reduces the maximum voltage swing at its drain, thereby limiting the maximum output power. These limitations may impede a conjugate match at the output and, hence, a power match is usually more appropriate.

The choice of the transformation network type is of major importance to achieve the desired matching. The main parameters of a matching network are the transformation and the loaded quality factors. The transformation factor (m) determines the transformed impedance, whereas the loaded quality factor (Q_0) determines the bandwidth of the transformation. The choice among the three most common types of matching network, L, π and T, is made upon these two properties.

L-sections can be either low-pass or high-pass. The inconvenience is that once the transformation factor is determined the quality factor is automatically set. π- and T-networks are more flexible allowing the choice of m and Q_0 independently. However, T-networks require two reactances in the series path to the antenna. In general, these reactances are inductors. Having two inductors in series in the output RF path is prohibitive in a CMOS design because of their low quality factor. In light of this and for other reasons that will be clarified in Sect. 6.2.1, a π-network was chosen for the output of the PA. More information on impedance matching network design can be found in [1].

6.2.1 The π-Matching Network

A typical π-matching network is shown in Fig. 6.2. In fact, it consists of two L-matching sections connected back to back and, hence, two transformations take place. The first section comprises C_2 and L_{1b} and promotes a downward transformation from R_L to R_t. (R_t is the resistance seen by L_{1a} when looking toward L_{1b}, as indicated in Fig. 6.2.) The second section consists of L_{1a} and C_1 and causes an upward transformation from R_t to R_{opt}. (R_{opt} is the resistance seen by the PA when looking toward the output matching network.) To each transformation we can

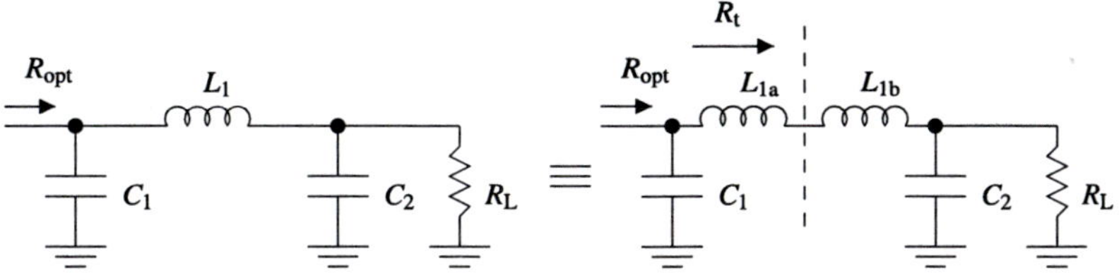

Fig. 6.2 Typical π-matching network

associate a quality factor:

$$Q_a = \frac{R_{opt}}{X_{C1}}, \tag{6.1}$$

$$Q_b = \frac{R_L}{X_{C2}}, \tag{6.2}$$

where X_{C1} and X_{C2} are the reactances of the capacitors C_1 and C_2.

For a given overall transformation factor, the components of the matching network can be determined, for instance, by choosing a value for Q_b. With Q_b and (6.2), the value of C_2 is determined. R_t is found by the downward transformation factor (m_b):

$$m_b = Q_b^2 + 1, \tag{6.3}$$

$$R_t = \frac{R_L}{m_b}. \tag{6.4}$$

The reactance of L_{1b} is determined through R_t and Q_b with

$$X_{L1b} = Q_b R_t. \tag{6.5}$$

The reactance of L_{1a} is determined by the upward transformation factor, R_t, and Q_a with

$$m_a = \frac{R_{opt}}{R_t}, \tag{6.6}$$

$$Q_a = \sqrt{m_a - 1}, \tag{6.7}$$

$$X_{L1a} = Q_a R_t, \tag{6.8}$$

whereas the reactance associated with C_1 can be found by replacing Q_a and R_{opt} in (6.1).

Inductor L_1 is the sum of L_{1a} and L_{1b} and, hence, all the three components are determined. An important parameter of a matching network is its loaded quality factor. According to Sun and Fidler in [27, (10)], its definition for a π-network is

$$Q_0 = \frac{Q_a + Q_b}{2}. \tag{6.9}$$

Table 6.1 π-matching network for a target $R_{\mathrm{opt}} = 18\ \Omega$. High-$Q_0$ case: $C_1 = 14.34$ pF and $C_2 = 8.68$ pF

freq (GHz)	L_1 (nH)	Q_0	R_{opt} (Ω)	X_{opt} (Ω)	Γ_{opt} (dB)
3.7	0.336	8.04	17.95	0.14	−47.43
3.8	0.319	8.25	17.90	0.32	−40.50
3.9	0.303	8.47	17.92	0.33	−40.39
4.0	0.288	8.69	17.93	0.34	−40.29
4.1	0.275	8.91	17.89	0.51	−36.66
4.2	0.262	9.13	17.90	0.52	−36.67
4.3	0.250	9.35	17.92	0.52	−36.70
4.4	0.239	9.57	17.93	0.52	−36.75
4.5	0.228	9.78	17.89	0.68	−34.28
4.6	0.219	10.00	17.90	0.68	−34.36
4.7	0.210	10.22	17.92	0.67	−34.46
4.8	0.201	10.44	17.87	0.84	−32.55
4.9	0.193	10.66	17.89	0.83	−32.66
5.0	0.185	10.88	17.90	0.82	−32.78
5.1	0.178	11.09	17.86	0.98	−31.20
5.2	0.171	11.31	17.87	0.97	−31.33

The loaded quality factor[1] is associated with the bandwidth of the network. Different expressions for Q_0 can be found in the literature [27, Appendix]. We chose (6.9) because it is in conformity with the definition of quality factor as the ratio between the maximum instantaneous energy stored and the energy dissipated per cycle.

The importance of Q_0 is twofold. First, we can verify if the network is designable according to the following criteria taken from [27]:

$$\begin{cases} Q_0 \geq \frac{1}{2}\sqrt{\dfrac{R_{\mathrm{opt}}}{R_{\mathrm{L}}} - 1} & \text{for } R_{\mathrm{opt}} \geq R_{\mathrm{L}}, \\[2ex] Q_0 \geq \frac{1}{2}\sqrt{\dfrac{R_{\mathrm{L}}}{R_{\mathrm{opt}}} - 1} & \text{for } R_{\mathrm{opt}} \leq R_{\mathrm{L}}. \end{cases} \tag{6.10}$$

Second, an interesting property of the π-matching network of Fig. 6.2 reveals that, for high Q_0, capacitors C_1 and C_2 can be kept constant while L_1 is changed to achieve the same transformation factor at different frequencies. This property can be noticed through the following example.

Consider that we would like to transform an antenna impedance of 50 Ω into an optimum resistance of 18 Ω in a frequency range from 3.7 to 5.2 GHz. If we choose a high-Q_0 π-matching network with $C_1 = 14.34$ pF and $C_2 = 8.68$ pF, Table 6.1 reveals that the value of R_{opt} can be kept very close to 18 Ω throughout the whole frequency range from 3.7 to 5.2 GHz. This is achieved by varying just L_1 whereas

---

[1]The term *loaded* describes the quality factor of a circuit element under loaded conditions [4, Chap. 2].

Table 6.2 π-matching network for a target $R_{\text{opt}} = 18\ \Omega$. Low-$Q_0$ case: $C_1 = 2.39$ pF and $C_2 = 1.84$ pF

freq (GHz)	L_1 (nH)	Q_0	R_{opt} (Ω)	X_{opt} (Ω)	Γ_{opt} (dB)
3.7	1.212	1.56	17.92	0	−53.2
3.8	1.169	1.62	18.07	0.30	−41.2
3.9	1.124	1.67	18.03	0.60	−35.6
4.0	1.082	1.72	18.00	0.88	−32.2
4.1	1.042	1.77	17.98	1.14	−29.9
4.2	1.004	1.82	17.95	1.40	−28.1
4.3	0.965	1.86	17.75	1.66	−26.5
4.4	0.931	1.91	17.73	1.90	−25.4
4.5	0.898	1.96	17.70	2.13	−24.4
4.6	0.867	2.01	17.68	2.36	−23.5
4.7	0.837	2.06	17.66	2.57	−22.7
4.8	0.807	2.11	17.47	2.79	−21.9
4.9	0.780	2.16	17.45	2.99	−21.3
5.0	0.755	2.21	17.44	3.19	−20.8
5.1	0.729	2.25	17.25	3.39	−20.1
5.2	0.706	2.30	17.23	3.57	−19.7

C_1 and C_2 are kept fixed. If we choose a low-Q_0 network, with $C_1 = 2.39$ pF and $C_2 = 1.84$ pF, Table 6.2 shows that the value of R_{opt} can still be kept close to 18 Ω throughout the whole frequency range from 3.7 to 5.2 GHz. This is again achieved by varying just L_1 whereas C_1 and C_2 are kept fixed.

For both cases, the transformed impedance Z_{opt} is close to the desired value of 18 Ω, but for the high-Q_0 case ($Q_0 > 8$) it is much closer than for the low-Q_0 case ($1.5 < Q_0 < 2.5$) as reveals the column Γ_{opt}—which presents a maximum value of -31.2 dB for high Q_0 and -19.7 dB for low Q_0. This column shows the reflection coefficient as if a port with a characteristic impedance (Z_0) of 18 Ω were connected to the output matching network in place of the power amplifier—refer to (6.11). Hence, the most negative the reflection coefficient is, the closer the transformed impedance is to the target resistance. Therefore, we can state that a higher Q_0 is preferred in the case in which the inductor is varied and the capacitors are kept constant.

$$\Gamma_{\text{opt}}\ (\text{dB}) = 20\log\left|\frac{Z_{\text{opt}} - Z_0}{Z_{\text{opt}} + Z_0}\right|. \tag{6.11}$$

Arrangements of inductors and capacitors, other than that of Fig. 6.2, can also be used to form narrowband π-matching networks as described in [1, Chap. 3]. In [7], a procedure for the design of a π-matching network resulting in a relatively wideband operation is discussed.

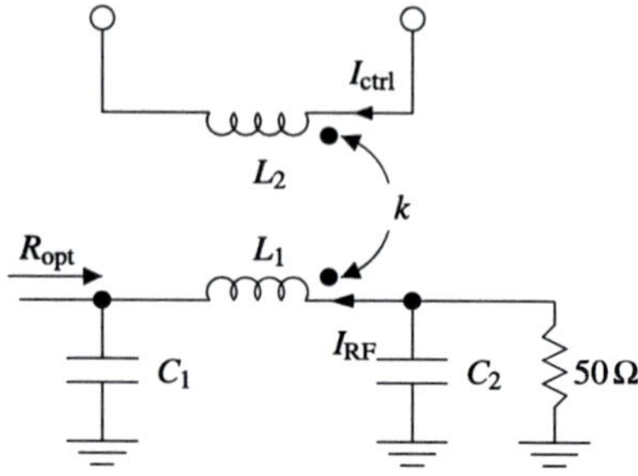

Fig. 6.3 Tunable π-matching network

6.2.1.1 Finite Q_u Inductors

An important non-ideality of an impedance matching network is the finite unloaded quality factor (Q_u) of the passive components. For the π-matching network, the inductor in the series path is the most critical component. We express the unloaded quality factor of an inductor as the ratio between its reactance and its parasitic series resistance:

$$Q_u = \frac{X_L}{R_{Ls}} = \frac{\omega L}{R_{Ls}}. \tag{6.12}$$

An inductor with finite Q_u in the series path of an output matching network reduces the output power of the PA. We distinguish two contributions:

1. The mismatch due to the presence of a resistor R_{Ls} in series with the inductor L_1 in Fig. 6.2. If this resistor is not taken into account correctly, it will produce an impedance mismatch which in turn will provoke a deviation of the impedance seen by the PA from its optimum value.
2. The power dissipation in the series resistor R_{Ls} due to the output RF current flowing from the PA to the antenna.

Hence, the quest for high-Q_u inductors in CMOS technology has its motivation also on the design of PAs.

6.2.2 Tunable Output Impedance Matching

The tunable output impedance matching network is based on the coupled-inductor technique introduced in Chap. 5. A π-type network using an inductor as the tunable element was chosen according to the analysis made in Sect. 6.2.1. A schematic of this network is shown in Fig. 6.3.

The terminals of L_2 are connected to a control circuit providing a current I_{ctrl} that satisfies (5.6) on p. 69. The design of the coupled inductors and the control circuit is the subject of Sect. 6.4.

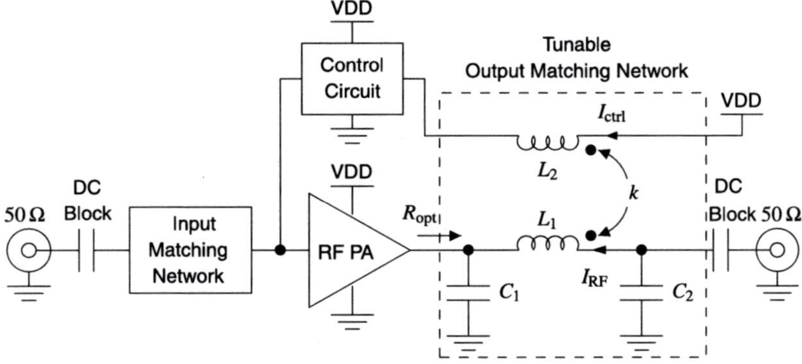

Fig. 6.4 Frequency-tunable RF power amplifier block diagram

6.3 Input Impedance Matching

The purpose of an input impedance matching network is to maximize the power transfer from the RF source to the PA input and, hence, to enhance the power gain of the amplifier [6, p. 20]. The main consequence of a bad input matching is a lower power gain. It will not, however, deteriorate the linearity performance although it can have an impact on the PAE if the power gain is less than 10 dB (refer to Sect. 2.2.2 on p. 6). If the power gain is high, the PAE is approximately equal to the drain efficiency. In such a case, a small reduction of the power gain caused by an input impedance mismatch will virtually not affect the PAE of the amplifier. For this reason, in the frequency-tunable PA herein described, we will focus on the design of a tunable output matching network, whereas the input matching will be kept fixed.

6.4 Frequency-Tunable RF Power Amplifier

The block diagram of the frequency-tunable PA is shown in Fig. 6.4. The main blocks are the RF PA, the fixed input matching network, the tunable output matching network, and the control circuit. The matching networks were analyzed in the previous sections. The other blocks as well as the complete system called frequency-tunable RF PA are treated in the sections that follow.

6.4.1 RF Power Amplifier

The RF power amplifier is essentially the same used in the dynamic supply PA whose design is described in Chap. 3. Its circuit is shown in Fig. 6.5. The components that are shown in the figure and the output matching network (not shown) are integrated. The input matching network was realized with off-chip components

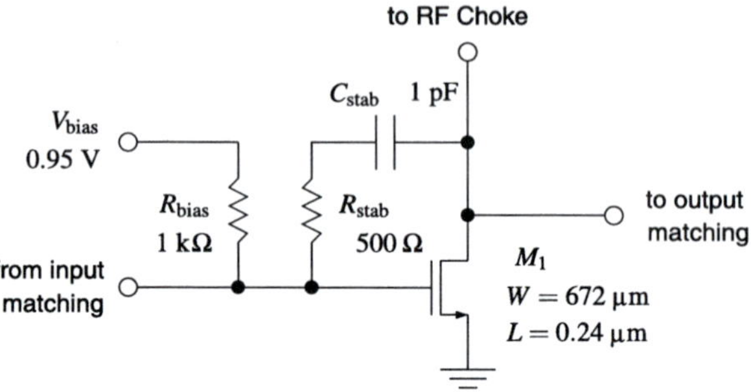

Fig. 6.5 Integrated CMOS RF power amplifier schematic

to provide more flexibility for adjustment. The RF choke is also an off-chip component. The RF PA was designed to operate in class A with a quiescent current of 87 mA and a supply voltage (VDD) of 2.5 V. Resistor $R_{stab} = 500\ \Omega$ and capacitor $C_{stab} = 1$ pF implement a shunt feedback that guarantees unconditional stability at all frequencies at the expense of reduced power gain. R_{stab} was implemented with a non-silicided polysilicon resistor and C_{stab} with a MIM capacitor. Although the minimum channel length for the CMOS technology used is 0.11 μm, 2.5 V devices are limited to 0.24 μm. This was the value chosen for M_1. It is obvious that choosing a smaller length would result in a better RF performance, but it would require a supply voltage of 1.2 V, which would limit the output power that could be attained with the same transistor size.

6.4.2 Coupled Inductors

In the previous chapter, the coupled-inductor technique was introduced as our choice to implement the tunable inductor for the output impedance matching network of the frequency-tunable PA. This choice was motivated by the property of the π-matching network for which keeping the capacitors fixed and varying only the inductor, the desired transformation can be achieved for different frequencies. Moreover, the possibility of Q enhancement makes the technique attractive due to the low quality factor of integrated inductors in CMOS technology that limits the attainable output power.

Some considerations about the ratio between the RF and control currents must be done before designing the coupled inductors. First, we recall some equations from last chapter:

$$\frac{I_{ctrl}}{I_{RF}} = \alpha + i\beta = r(\cos\phi + i\sin\phi), \tag{6.13}$$

$$L_{eq} = L(1 + \alpha k) = L(1 + kr\cos\phi), \tag{6.14}$$

$$R_{eq} = -\omega k\beta L + R_{Ls} = -\omega k L r \sin\phi + R_{Ls}. \tag{6.15}$$

The domains of α and β values can be determined by checking the corresponding equivalent inductance and resistance. Figure 6.6 depicts these domains for different

Fig. 6.6 ϕ domain in
quadrants

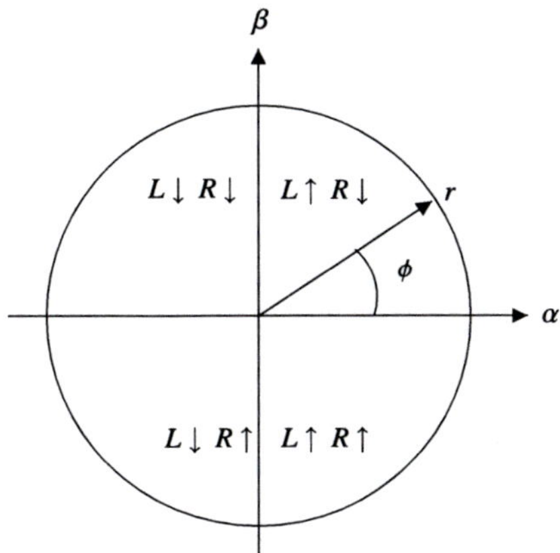

values of the phase ϕ between the two currents. The Cartesian plane is divided in
four quadrants and it can be seen that the region of interest encloses the first and
second quadrants because the corresponding equivalent resistance is lower than the
original value R_{Ls}—Q enhancement—and the inductance can be greater or lower
than the original value L—inductance tuning. Operation in these quadrants imposes
that $\beta > 0$ or $0° < \phi < 180°$.

The value of r is limited by the efficiency of the circuit. The higher r is, the
higher the control current is. Therefore, the target is $r < 1$ and as close to zero as
possible.

6.4.2.1 Design of the Coupled Inductors

The design of coupled inductors (transformers) is a complex subject treated in several articles, thesis, and textbooks like [9, 12, 15, 19, 25]. An analysis of the various
types of integrated transformers can be found in [12, 15]. To choose the type of the
transformer to design the coupled inductors, we identify two main requirements:
high coupling factor and 4-terminal configuration.

The following types of transformers were analyzed:

1. *Tapped*: consists of two concentric windings for which only one turn of each
   winding is adjacent to a turn of the other winding. This limits the coupling factor
   to values between 0.3 and 0.5. It is suitable for 3-terminal applications.
2. *Interleaved*: also known as Frlan transformer [12], because it was first proposed
   by Frlan in [8]. It consists of two interwound coils implemented with the same
   metal layer. Coupling factors from 0.6 to 0.8 can be achieved [12, 15] and it is
   appropriate to 4-terminal applications. Because of the distance between the turns
   of the same winding, the self-inductance is usually low.
3. *Stacked*: also known as overlay or Finlay transformer [12]. It is composed of
   two windings that are implemented each in a different metal layer. By exploiting
   both lateral and vertical magnetic coupling, this type of transformer achieves the

highest self-inductance and coupling factor ($k \approx 0.9$). The drawback is the high capacitance between winding terminals which translates to a lower SRF.

4. *Stacked Interleaved*: by using windings stacked in two different metal layers, but with both windings being interleaved in each layer, coupling factors of the order of 0.7 can be achieved [15]. The advantage over the interleaved topology is the increased inductance density.

5. *Rabjohn*: first proposed by Rabjohn in [25], this transformer has interwound windings and exhibits perfect symmetry when center tapped and, hence, is suitable for application as a balun [12].

A decision for an interleaved transformer is justified by its suitability for application in a 4-terminal configuration and because of its relatively high coupling factor. Although the stacked configuration offers a higher coupling factor, it was not chosen because of a possible lower SRF.

Interleaved Transformer Design and Simulation   The design of the transformer was an iterative process using Matlab [14], VeloceRF [11], and ADS Momentum [2]. The geometrical parameters for the layout were obtained using the Matlab function *indspi* from the *MLib* toolbox developed by Pisani in [22, 23]. The formulae used in this function are based on the works reported in [3, 16]. The geometrical parameters were then entered in VeloceRF which generated the layout for the transformer. This layout was then imported into ADS Momentum where a 2.5D planar electromagnetic simulation was performed.

In order to generate the initial geometrical parameters for the transformer, we calculated the mutual inductance using the method described by Mohan in [15]. In this method, two inductances are associated to the transformer. The first one, $L = L_1 = L_2$, is the inductance of the primary and secondary windings. The second one, L_T, is the inductance of a single spiral containing all the segments of both windings. The mutual inductance can be written as

$$M = \frac{L_T - L_1 - L_2}{2}. \tag{6.16}$$

The coupling factor (k) can be directly determined from (6.16):

$$k = \frac{M}{\sqrt{L_1 L_2}}. \tag{6.17}$$

Table 6.3 shows the parameters obtained using the *indspi* function considering $L_1 = L_2 = 0.8$ nH. The definition of the geometrical parameters is provided in Fig. 6.7. The choice of the value for the self-inductances L_1 and L_2 depends on the design of the control circuit and is treated in Sect. 6.4.3. It is important to note that the coupling factor given by ADS Momentum was not available at the time the circuit was designed. This result was obtained after applying the model described in Sect. 7.3 on p. 104. This investigation was carried out to find out the causes of the malfunctioning of the circuit. Hence, for the design, we considered $k = 0.7$, which is lower than that obtained with (6.16), thereby providing a safety margin.

The layout of the interleaved transformer was generated using VeloceRF with the geometrical parameters of Table 6.3. A Patterned Ground Shield (PGS) and

Table 6.3 Interleaved square transformer parameters

Parameter	Transformer	L	L_T
Outer diameter (d_out)	197 µm	180 µm	197 µm
Inner diameter (d_in)	99 µm	116 µm	99 µm
No. of turns (n)	1.5	1.5	3
Track width (w)	15 µm	15 µm	15 µm
Track spacing (s)	2 µm	19 µm	2 µm
No. of segments	12	6	12
Metal layer	top	top	top
Matlab L	–	0.61 nH	2.03 nH
Momentum L @ 5.2 GHz	–	0.75 nH	2.81 nH
Momentum R_Ls @ 5.2 GHz	–	2.15 Ω	15.5 Ω
Momentum Q_u @ 5.2 GHz	–	11.4	5.9
k acc. to (6.16)	0.87	–	–
Momentum k @ 5.2 GHz	0.39	–	–

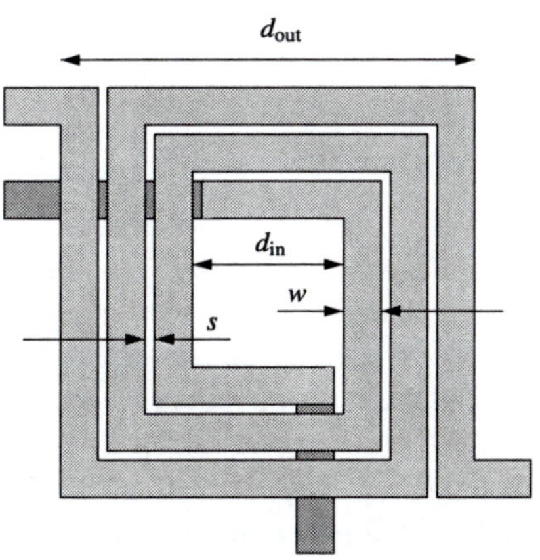

Fig. 6.7 Interleaved square transformer showing geometrical parameters

dummy structures required by the technology design rules were added to the layout, as shown in Fig. 6.8, and simulated using ADS Momentum. These simulation results are also shown in Table 6.3. The PGS was implemented using Metal 1 and its function is to increase the quality factor of the inductor by decreasing the substrate losses. The substrate losses can be reduced by decreasing the substrate resistance as if the substrate were a short circuit to ground. This can be achieved by inserting a ground plane between the spiral and the substrate. A PGS, instead of a solid ground shield, is more effective because it prevents the induction of loop currents in the ground plane which would reduce the overall inductance of the spiral [28]. Another possibility for reduction of the substrate losses is to increase the substrate resistance so that the substrate mimics an open circuit. This can be achieved with high-resistivity substrates [24]. However, this is not a common option in standard CMOS technologies.

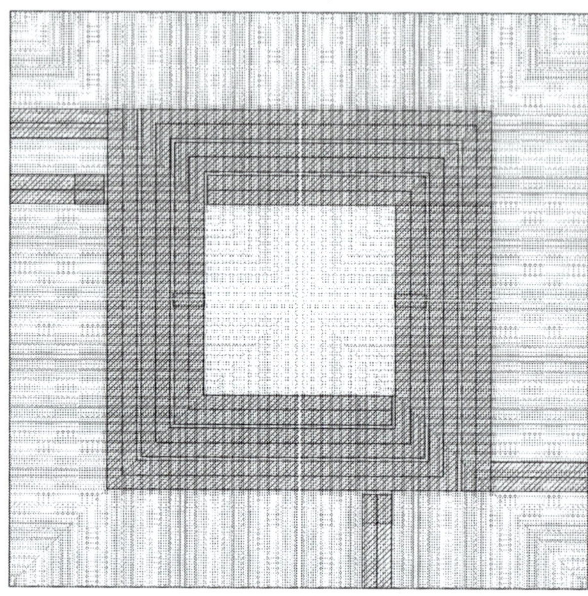

Fig. 6.8 Layout of the interleaved coupled inductors with patterned ground shield and dummy structures

6.4.3 Control Circuit

As shown in the block diagram of Fig. 6.4, the control circuit forces a current I_{ctrl} in the secondary winding (control winding) of the coupled inductors which depends on the input RF signal. I_{ctrl} is related to the current I_{RF} flowing in the primary winding (RF winding) of the coupled inductors because I_{RF} itself depends on the input RF signal. Figure 6.9 shows a possible implementation of the control circuit.

The circuit of Fig. 6.9 is designed using the Signal-Flow Graph (SFG) technique [13]. The SFG proves useful for representing circuit equations and deriving its transfer functions, especially for circuits with multiple loops. Examples of SFGs in circuits with transformers can be found in [5, 10].

We also use the Driving-Point Impedance (DPI) concept [20, 21, 26]. Together with the SFG technique, it provides a systematic procedure (known as the DPI/SFG analysis [21, Chap. 4]) for deriving the transfer functions of a circuit.

6.4.3.1 Analysis of the Control Circuit

The small-signal model of the control circuit of Fig. 6.9 is depicted in Fig. 6.10. Two transconductances are used to establish a relationship between I_{ctrl} and I_{RF}. We can further simplify the analysis by replacing the coupled inductors by their two-port network representation shown in Fig. 6.11.

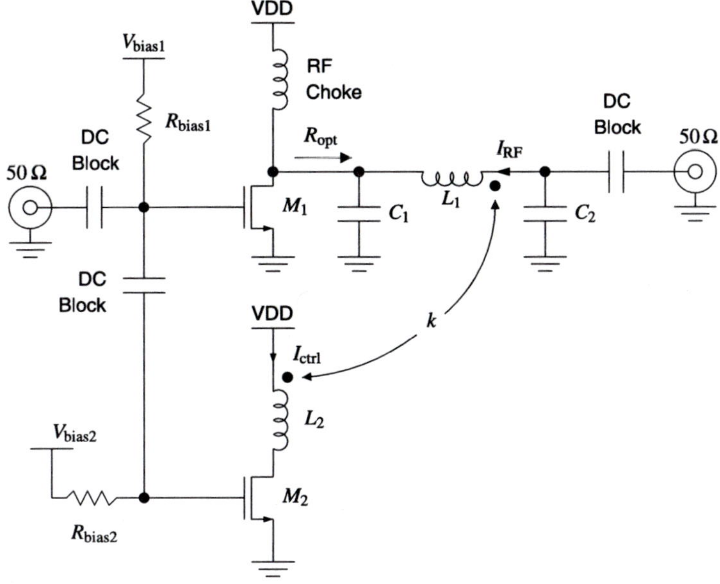

Fig. 6.9 Simplified frequency-tunable RF PA circuit

Fig. 6.10 Small-signal model for the circuit of Fig. 6.9

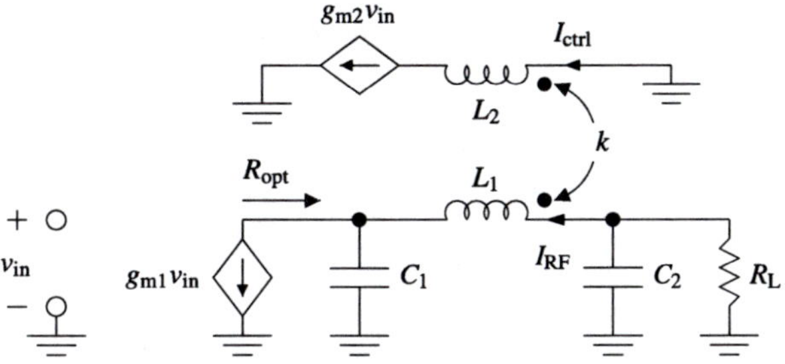

The equations for the coupled inductors can be arranged in a Z-parameter matrix representation:

$$\begin{bmatrix} V_1 \\ V_2 \end{bmatrix} = \begin{bmatrix} \overbrace{iX_{L1}}^{z_{11}} & \overbrace{iX_M}^{z_{12}} \\ \underbrace{iX_M}_{z_{21}} & \underbrace{iX_{L2}}_{z_{22}} \end{bmatrix} \begin{bmatrix} I_1 \\ I_2 \end{bmatrix}. \tag{6.18}$$

The small-signal equivalent circuit of Fig. 6.12 is obtained by replacing the coupled inductors of Fig 6.10 by the Z-parameter model of Fig. 6.11. In order to apply the SFG/DPI method, we recognize that there are $n = 6$ nodes in this circuit with the ground node excluded. There are also $p = 3$ voltage sources (V_{in}, $iX_M I_{\text{RF}}$, and $iX_M I_{\text{ctrl}}$). According to [21, p. 82], $n - p = 3$ auxiliary voltage sources must be added to the circuit (V_{out}, V_{x}, and V_{y}).

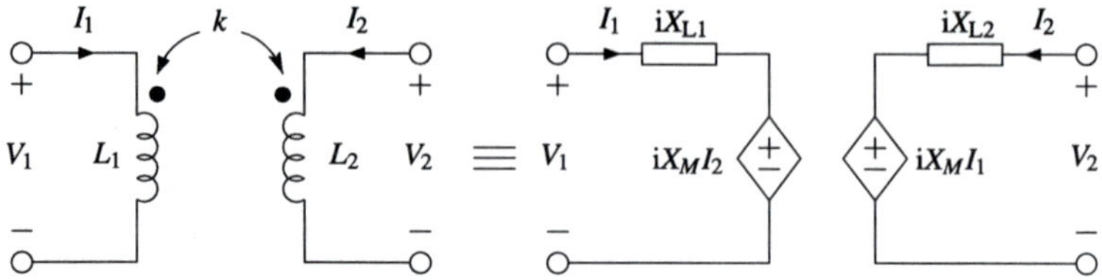

Fig. 6.11 Z-parameter transformer model

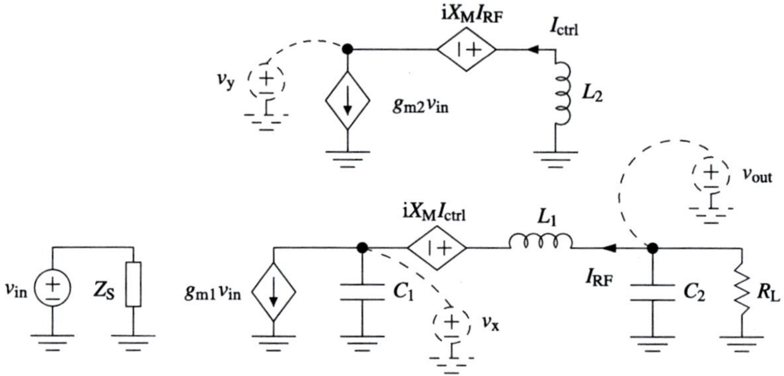

Fig. 6.12 Small-signal model used to derive the control circuit SFG. Auxiliary voltage sources are also shown (drawn with *dashed lines*)

Next, we must write the equations for the driving-point impedances and short-circuit currents for each node with an auxiliary voltage source. To calculate the DPI of a node, the impedance seen by the auxiliary voltage source is determined by zeroing all the other sources in the circuit—including the other auxiliary voltage sources. With $L_1 = L_2 = L$, the following impedances are obtained:

$$\begin{cases} Z_{\text{DPx}} = \frac{sL}{1+s^2LC_1}, \\ Z_{\text{DPy}} = sL, \\ Z_{\text{DPout}} = \frac{sR_LL}{R_L+sL+s^2LC_2}. \end{cases} \tag{6.19}$$

The short-circuit current of a node is obtained by connecting the node to ground, setting to zero all but one source and determining the current entering the node. For each node, every source in the circuit must be taken into account. The following short-circuit currents are obtained:

$$\begin{cases} i_{\text{SCx}} = -g_{m1}v_{\text{in}} - \frac{iX_M}{sL}I_{\text{ctrl}} + \frac{1}{sL}v_{\text{out}}, \\ i_{\text{SCy}} = -g_{m2}v_{\text{in}} - \frac{iX_M}{sL}I_{\text{RF}}, \\ i_{\text{SCout}} = \frac{iX_M}{sL}I_{\text{ctrl}} + \frac{1}{sL}v_{\text{x}}. \end{cases} \tag{6.20}$$

We recognize that

$$I_{\text{ctrl}} = g_{m2}v_{\text{in}}, \tag{6.21}$$

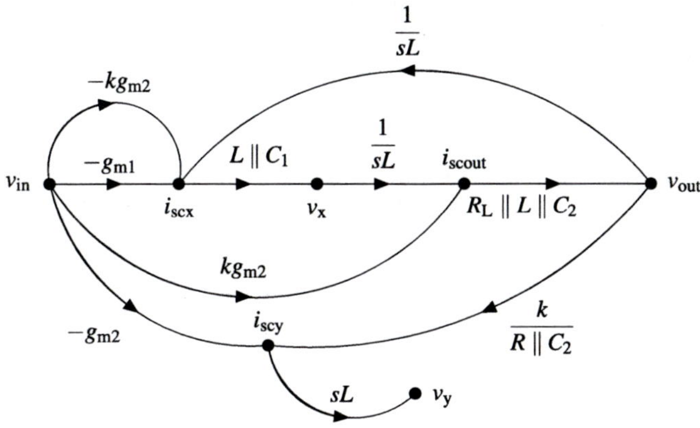

Fig. 6.13 Signal-flow graph of the circuit of Fig. 6.9

$$I_{RF} = -\frac{1 + s R_L C_2}{R_L} v_{out}, \tag{6.22}$$

$$\frac{i X_M}{s L} = k, \tag{6.23}$$

and, hence, (6.20) can be rewritten as

$$\begin{cases} i_{SCx} = -g_{m1} v_{in} - k g_{m2} v_{in} + \frac{1}{sL} v_{out}, \\ i_{SCy} = -g_{m2} v_{in} + k \frac{1 + s R_L C_2}{R_L} v_{out}, \\ i_{SCout} = k g_{m2} v_{in} + \frac{1}{sL} v_x. \end{cases} \tag{6.24}$$

The voltages of the nodes in the circuit with an associated auxiliary source are determined by multiplying the driving-point impedance by the short-circuit current—$v_n = Z_{DPn} \times i_{SCn}$. The voltages of the remaining nodes are determined by the sum of the auxiliary and existing voltage sources.

Once the driving-point impedances and short-circuit currents have been determined, the SFG of Fig 6.13 can be drawn—the symbol "∥" represents a parallel circuit combination. From this graph it is possible to determine the transfer function of the circuit:

$$H(s) = \frac{v_{out}}{v_{in}}. \tag{6.25}$$

However, we are interested in obtaining the ratio between the control and RF currents. Division of (6.21) by (6.22) yields the desired current ratio as a function of $H(s)$:

$$\frac{I_{ctrl}}{I_{RF}} = -g_{m2} \frac{R_L}{1 + s R_L C_2} H(s)^{-1}. \tag{6.26}$$

Graph algebra and graph reduction techniques can be used to simplify the SFG of Fig. 6.13 and determine $H(s)$. Alternatively, an SFG analysis tool for Matlab called *flow_tf* [17, 18] can generate automatically the transfer function between two nodes of an SFG from the circuit equations.

Table 6.4 Script *tunable_ratio.flw* for the *flow_tf* tool

```
vin iscx -(gm1+k*gm2)

vin iscy -gm2

iscx vx (s*L)/(1+s^2*L*C1)

iscy vy s*L

vx iscout 1/(s*L)

vin iscout k*gm2

iscout vout (s*RL*L)/(RL+s*L+RL*L*C2*s^2)

vout iscy k*(1/RL+s*C2)

vout iscx 1/(s*L)

.pre syms s gm1 gm2 RL C1 C2 L k

.tf vin vout
```

Table 6.5 Matlab commands used with the *flow_tf* tool

```
H=flow_tf('tunable_ratio.flw');

syms s gm1 gm2 RL C1 C2 L k;

Iratio=-gm2*(RL/(1+s*RL*C2))*(1/H.sym{1});

pretty(Iratio);
```

Replacing the expression for $H(s)$ calculated with the *flow_tf* tool into (6.26) yields

$$\frac{I_{\text{ctrl}}}{I_{\text{RF}}} = -\frac{g_{\text{m2}}[1 + sR_{\text{L}}(C_1 + C_2) + s^2 LC_1 + s^3 R_{\text{L}} LC_1 C_2]}{(1 + sR_{\text{L}}C_2)(-g_{\text{m1}} + s^2 g_{\text{m2}} kLC_1)}. \tag{6.27}$$

The script used with *flow_tf* is shown in Table 6.4 and the Matlab commands to run the script and calculate the final current ratio are shown in Table 6.5.

6.4.3.2 Design of the Control Circuit

Equation (6.27) provides a relationship between the components of the circuit of Fig. 6.9 and the current ratio (I_{ratio}) between I_{ctrl} and I_{RF}. It was seen that the phase and amplitude of I_{ratio} determines the effective inductance and resistance seen by the RF circuit between the two terminals of L_1. Agilent's ADS simulator provides a direct calculation of the current ratio.

In order to establish a design procedure for the control circuit, we identify three main points of the problem:

1. Determination of the tuning range required for the inductance.
2. Determination of α and β (or r and ϕ).
3. Feasibility of the circuit for the required variation of α and β.

For the first point, the knowledge of five variables is required: R_{opt}, C_1, C_2, the nominal value of L_1, and its unloaded quality factor. The determination of these

Table 6.6 π-matching network component values at 3.7 GHz for $R_{\mathrm{opt}} = 18\ \Omega$

C_1	L_1	C_2	Q_0
0.61	1.20	1.20	0.83
3.85	0.99	2.58	2.31
5.41	0.79	3.44	3.13
6.91	0.65	4.30	3.95
8.39	0.55	5.16	4.75
9.85	0.48	6.02	5.56
11.31	0.42	6.88	6.37
12.76	0.38	7.74	7.17
14.21	0.34	8.60	7.97

Table 6.7 π-matching network component values at 5.2 GHz for $R_{\mathrm{opt}} = 18\ \Omega$

C_1	L_1	C_2	Q_0
1.52	0.89	1.22	1.45
2.74	0.71	1.84	2.31
3.85	0.56	2.45	3.13
4.92	0.46	3.06	3.95
5.97	0.39	3.67	4.75
6.91	0.35	4.22	5.48
8.05	0.30	4.90	6.37
9.08	0.27	5.51	7.17
9.85	0.25	5.97	7.77

variables is an iterative process requiring some constraints to be established. The optimum resistance is first determined by the loadline method [6, Chap. 2]. This resistance is used to calculate the components of the output π-matching network. The two frequencies of interest are 3.7 and 5.2 GHz. For each frequency, the values of L_1 and Q_u will be different, but the values of C_1 and C_2 will remain constant.

The value of L_1 is limited by the technology and cannot be lower than 0.35 nH. There are also constraints regarding the designability of the π-matching network [27]. The values of C_1 and C_2 must be chosen carefully so that they provide the desired transformation for the two frequencies of interest—knowing that C_1 and C_2 do not change with frequency. Tables 6.6 and 6.7 show the values required for C_1, L_1, and C_2 to transform the 50 Ω antenna impedance into the required optimum resistance of 18 Ω at 3.7 and 5.2 GHz. The values are given for different Q_0. As already mentioned earlier in this chapter, the required values of C_1 and C_2 vary less with frequency for high Q_0.

Three cases of interest are highlighted. With $Q_0 = 7.77$ at 5.2 GHz (*solid line* linking rows of the two tables), the values required are $C_1 = 9.85$ pF, $L_1 = 0.25$ nH, and $C_2 = 5.97$ pF. At 3.7 GHz, the values required are $C_1 = 9.85$ pF, $L_1 = 0.48$ nH, and $C_2 = 6.02$ pF, which shows that the required values at 3.7 and 5.2 GHz are very close for C_1 and C_2. This means that by varying the inductor from 0.48 nH at

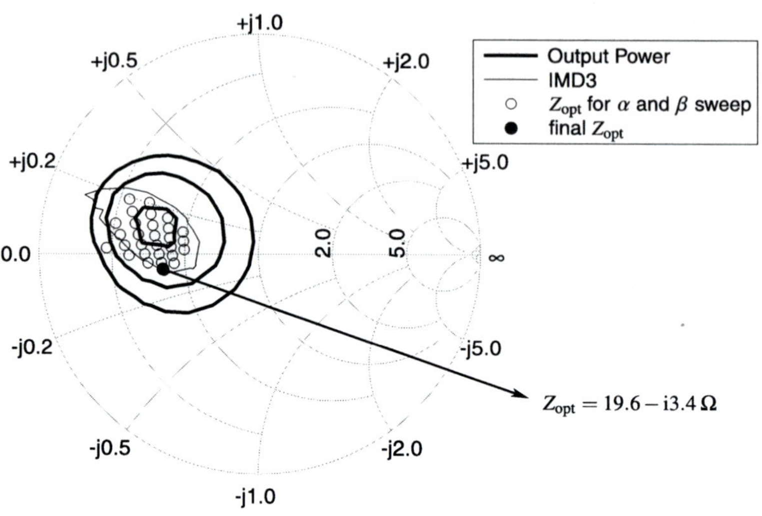

Fig. 6.14 Load-pull contours at 3.7 GHz

3.7 GHz to 0.25 nH at 5.2 GHz, the capacitors can be kept the same and the desired transformation will be achieved. However, the values for the inductor are too low to be implemented in practice and higher values lead to lower Q_0 and, hence, to an error in the transformation factor when just the inductor is varied.

In our case, we choose a higher value of nominal inductance, 0.8 nH, and to work respecting the *dotted line* linking the first rows of the Tables 6.6 and 6.7 (final values of C_1 and C_2 were 0.9 and 1 pF, respectively). For this case, an inductor tuning range from 0.9 to 1.2 nH (TR = 28%) is required. This corresponds to an α—see (5.11) on p. 70—between 0.235 and 0.66 (if we consider $k = 0.7$). Because of the low value of Q_0, some errors in the transformation factor are expected. We use, however, load-pull techniques to check if the transformation achieved is within an acceptable range.

To verify if the desired PA performance is reached with the values chosen for the capacitors, a load-pull test using 2-tone simulation is performed at 3.7 and 5.2 GHz. The real and imaginary parts of the impedance presented at the output of the PA are swept around R_{opt}. In our case, $R_{opt} = 18\,\Omega$ for an output power of 16 dBm (40 mW) at a maximum IMD3 of -35 dBc. This yields the contours of output power and IMD3 shown in the Smith charts of Figs. 6.14 and 6.15. The thick contours indicate the output power, whereas the thin ones indicate the IMD3 level. Three contours for output power are plotted in the figure. The inner is for $P_{out} > 16.5$ dBm (45 mW), the intermediate for $P_{out} > 16$ dBm (40 mW), and the outer for $P_{out} > 15.5$ (35.5 mW). The single thin contour encloses regions for which the IMD3 is lower than -35 dBc. The purpose is to find which values of α and β result in a transformed resistance corresponding to points inside the contours of Figs. 6.14 and 6.15. The regions outside the contours correspond to output power levels lower than 15.5 dBm (35.5 mW) or IMD3 levels greater than -35 dBc.

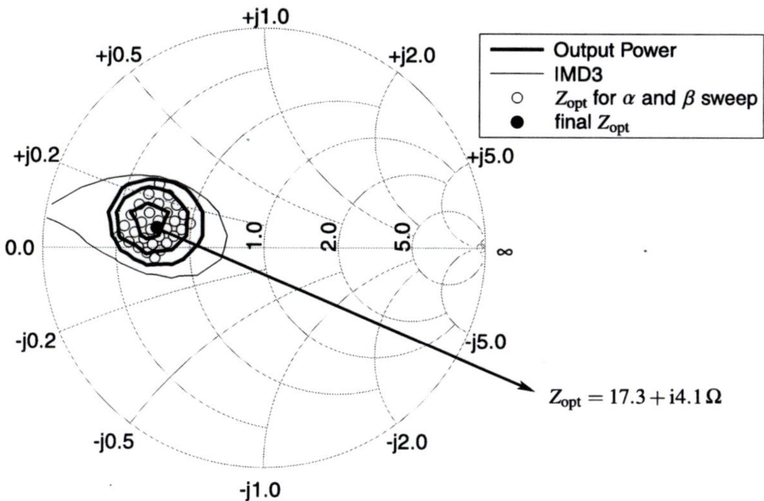

Fig. 6.15 Load-pull contours at 5.2 GHz

Fig. 6.16 π-matching network used to simulate the variation of α and β

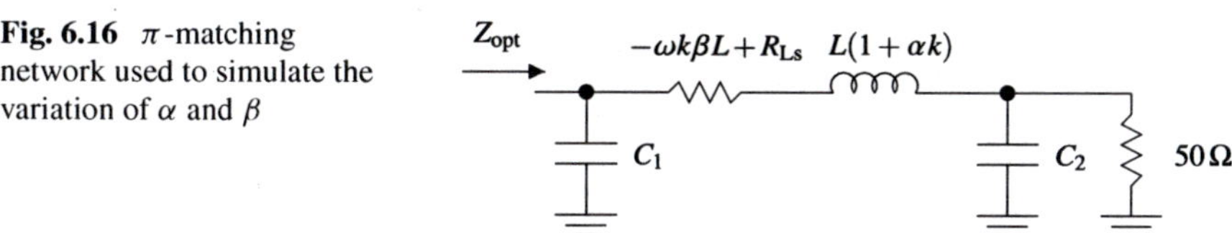

The output π-matching network is then simulated at 3.7 and 5.2 GHz. To make it possible to sweep the values of α and β, the tunable inductor is replaced by its equivalent resistance and inductance, whose expressions are given in (5.9) and (5.11) on p. 69, as shown in Fig. 6.16. The ranges of α and β are adjusted so that the transformed impedance Z_{opt} falls inside the IMD3 and output power contours as depicted in Figs. 6.14 and 6.15. For 3.7 GHz, the ranges are $0.6 < \alpha < 1.4$ and $0.2 < \beta < 1.1$, and for 5.2 GHz they are $0 < \alpha < 0.5$ and $0 < \beta < 0.4$.

To verify if the values of α and β can be achieved with the conceptual circuit of Fig. 6.9, an AC simulation is performed sweeping the bias voltage V_{bias2} of transistor M_2 and its width W_2. The main results from this simulation are summarized in Table 6.8, where I_{ctrl_q} is the quiescent control current. Therefore, we conclude that the conceptual circuit can be used.

6.4.4 Complete System

The complete system is shown in Fig. 6.17. A cascode structure (M_2–M_3) replaces the simple transistor M_2 of Fig. 6.9. This decreases the feedback from the node v_y (see Fig. 6.12) to the input of the circuit. The input impedance matching network comprising L_3 and C_3 in an L-type low-pass matching section is also shown.

Table 6.8 Summary of the load-pull test results

Parameter	3.7 GHz	5.2 GHz
W_2 (µm)	180	80
V_{bias2} (V)	0.7	0.7
I_{ctrl_q} (mA)	16.5	7.5
α	0.66	0.25
β	0.33	0.17
L_{eq} (nH)	1.15	0.93
R_{eq} (Ω)	-1.32	-0.15
Z_{opt} (Ω)	$19.6 - i3.4$	$17.3 + i4.1$

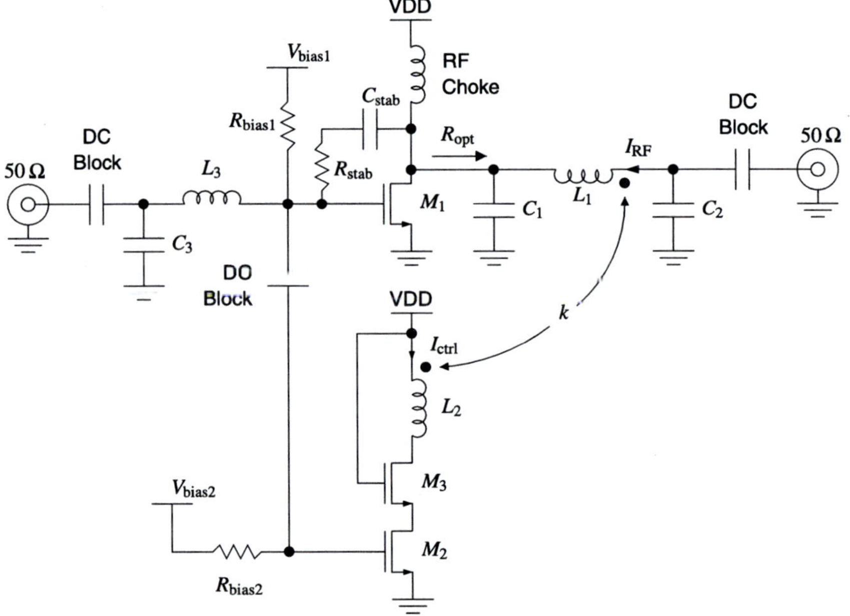

Fig. 6.17 Complete circuit for the frequency-tunable RF PA

Figures 6.18 and 6.20 show the IMD3 results for a 2-tone test using envelope simulation at 3.7 and 5.2 GHz. These figures also show the -35 dBc IMD3 target for a maximum acceptable distortion (*dashed line*). Figures 6.19 and 6.21 show the PAE results at 3.7 and 5.2 GHz. The values at maximum linear output power are also highlighted. These results are summarized in Table 6.9. For the conventional PA, the control circuit is disabled and does not consume any power. The coupled inductors are eliminated and L_1 is replaced by an on-chip inductor of the required value ($L = 0.92$ nH, $Q_u = 8.8$) so that $R_{opt} = 18\ \Omega$ is obtained at 5.2 GHz.

At 3.7 GHz and at a -35 dBc IMD3 level, the tunable PA presents an output power of 14.7 dBm (29.5 mW) and a PAE of 8.8% against 13 dBm (20 mW) and

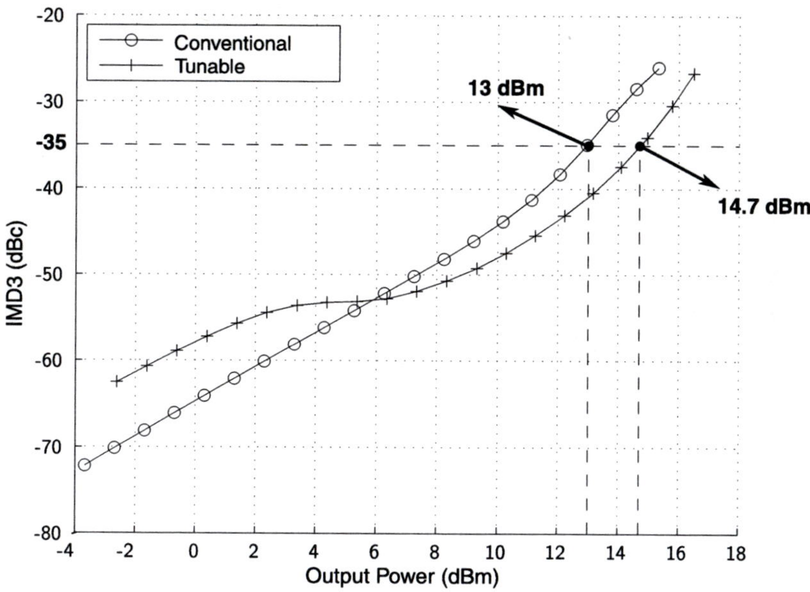

Fig. 6.18 Simulated IMD3 at 3.7 GHz. IMD3 limit of −35 dBc is also shown (*dashed line*)

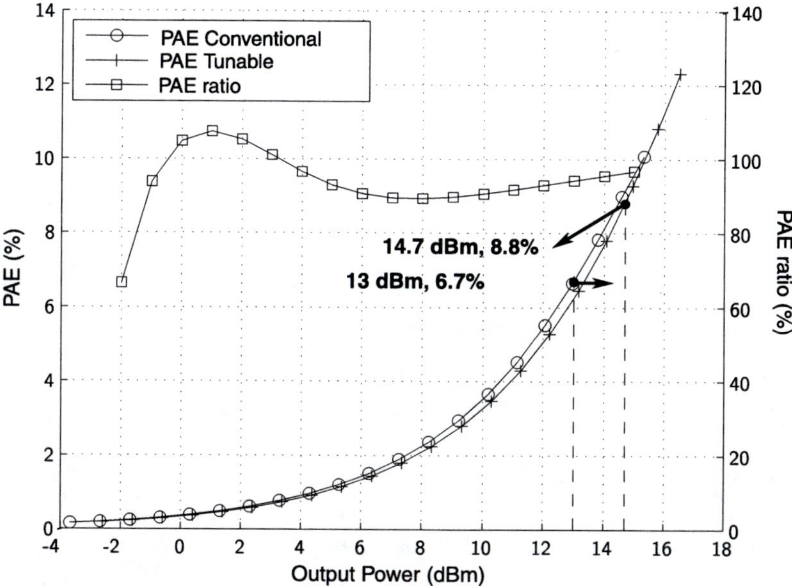

Fig. 6.19 Simulated PAE at 3.7 GHz

6.7% of its conventional counterpart. At 5.2 GHz, the tunable PA attains 14.8 dBm (30.2 mW) and 11.5%, whereas the conventional PA attains 13.2 dBm (20.9 mW)

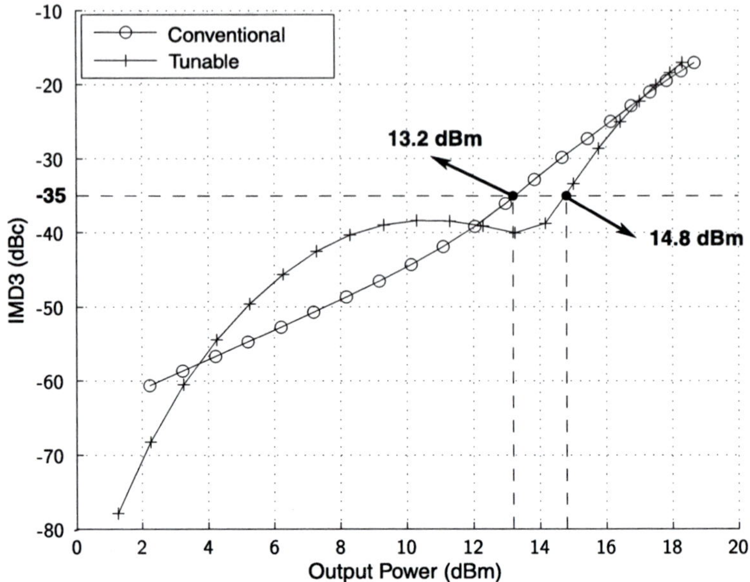

Fig. 6.20 Simulated IMD3 at 5.2 GHz. IMD3 limit of −35 dBc is also shown (*dashed line*)

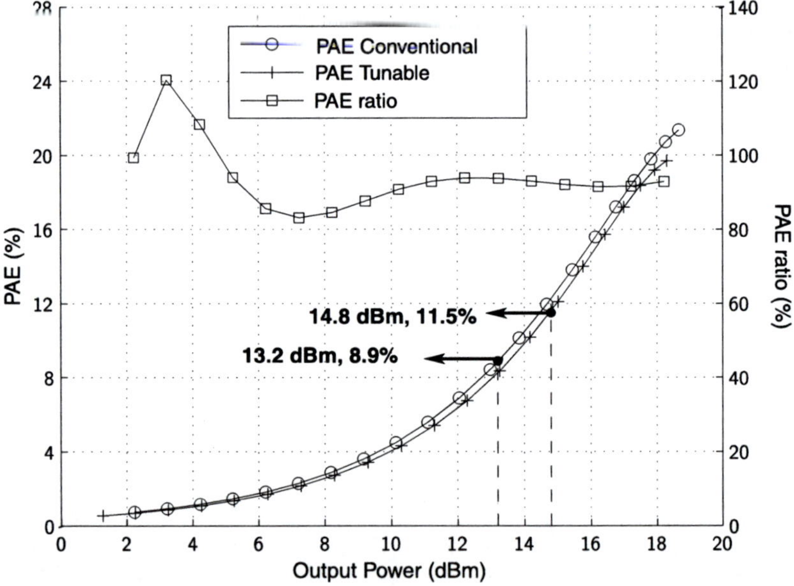

Fig. 6.21 Simulated PAE at 5.2 GHz

and 8.9%. These values reveal that the performance of the PA is improved using the tunable circuit when we consider efficiency and output power at a given distortion level.

Table 6.9 Simulated 2-tone output power and PAE at −35 dBc IMD3

2-tone	3.7 GHz		5.2 GHz	
	Conv.	Tun.	Conv.	Tun.
Pout (dBm)	13	14.7	13.2	14.8
PAE (%)	6.7	8.8	8.9	11.5

However, if we do not look at the distortion and only compare the efficiency at the same output power level, the conventional PA presents a slightly better PAE. This difference in efficiency can be measured by the ratio between the PAE of the tunable and that of the conventional PA shown in Figs. 6.19 and 6.21. A PAE ratio of 100% corresponds to equal efficiencies. The difference is large only at very low output power levels and is very small elsewhere. Although, the consumption of the control circuit is included in the PAE calculation, the small difference in efficiencies can be explained by the fact that, due to the Q enhancement in the frequency-tunable PA, the parasitic series resistance of the inductor is decreased and, consequently, the power dissipated on it is lower.

These results indicate that the coupled-inductor technique can be used in a frequency-tunable PA for dual-band operation at 3.7 and 5.2 GHz (TR $= 34\%$).

6.5 Conclusion

This chapter presented the design of a frequency-tunable CMOS RF power amplifier. It showed that a π-type output impedance matching network is suitable if only the inductor in the series path is adjusted as frequency changes. The design of the coupled inductors and the control circuit used to implement this variable inductance was presented. The PA design was briefly discussed. Its design follows that of the PA used with the dynamic supply in Chap. 3. Simulation results of the overall system demonstrated that a better performance can be achieved when operating in two different frequency bands if the frequency-tunable capability is implemented. In our analysis, the 3.7 and 5.2 GHz frequency bands were considered.

References

1. Abrie PLD (1985) The Design of Impedance-Matching Networks for Radio-Frequency and Microwave Amplifiers. Artech House, Dedham
2. Agilent (2010) Momentum. URL http://eesof.tm.agilent.com/products/momentum_main.html
3. Biondi T, Scuderi A, Ragonese E, Palmisano G (2005) Sub-nH inductor modeling for RFIC design. IEEE Microw Wirel Compon Lett 15(12):922–924
4. Bowick C (1982) RF Circuit Design. Newnes, Burlington
5. Cassan DJ, Long JR (2003) A 1-V transformer-feedback low-noise amplifier for 5-GHz wireless LAN in 0.18-μm CMOS. IEEE J Solid-State Circ 38(3):427–435

6. Cripps SC (2006) RF Power Amplifiers for Wireless Communications, 2nd edn. Artech House, Norwood
7. Dutta Roy SC (2000) Exact solution of a bandpass matching problem. Circ Systems Signal Process 19(1):59–69
8. Frlan E, Meszaros S, Cuhaci M, Wight JS (1989) Computer aided design of square spiral transformers and inductors. In: IEEE MTT-S Int Microw Symp Dig (IMS'89), Long Beach, CA, vol 2, pp 661–664
9. Gan H (2006) On-chip transformer modeling, characterization, and applications in power and low noise amplifiers. PhD thesis, Stanford University, Stanford, CA
10. Gee WA (2005) CMOS integrated LC Q-enhanced RF filters for wireless receivers. PhD thesis, Georgia Institute of Technology, Atlanta, GA
11. Helic (2010) VeloceRF. URL http://www.helic.com/index.php/velocerf/velocerf
12. Long JR (2000) Monolithic transformers for silicon RF IC design. IEEE J Solid-State Circ 35(9):1368–1382
13. Mason SJ (1953) Feedback theory—some properties of signal flow graphs. Proc IRE 41(9): 1144–1156
14. MathWorks (2010) Matlab. URL http://www.mathworks.com/
15. Mohan SS (1999) The design, modeling and optimization of on-chip inductor and transformer circuits. PhD thesis, Stanford University, Stanford. URL http://smirc.stanford.edu/papers/Thesis-mohan.pdf
16. Mohan SS, del Mar Hershenson M, Boyd SP, Lee TH (1999) Simple accurate expressions for planar spiral inductances. IEEE J Solid-State Circ 34(10):1419–1424
17. Neitola M, Rahkonen T (2005) A fully automated flowgraph analysis tool for Matlab. In: Proc Eur Conf Circuit Theory Des (ECCTD'05), Cork, Ireland, vol 1, pp 185–188
18. Neitola M, Rahkonen T (2005) A fully automated flowgraph analysis tool for MATLAB. URL http://www.mathworks.com/matlabcentral/fileexchange
19. Niknejad AM, Meyer RG (2000) Design, Simulation and Applications of Inductors and Transformers for Si RF ICs. Kluwer Acad, Boston
20. Ochoa JA (1998) A systematic approach to the analysis of general and feedback circuits and systems using signal flow graphs and driving-point impedance. IEEE Trans Circ Syst II 45(2):187–195
21. Phang K (2001) CMOS optical preamplifier design using graphical circuit analysis. PhD thesis, University of Toronto, Toronto, Canada
22. Pisani MB (2006) MLib—RF analysis and passive design toolbox. Matlab Toolbox. URL http://legwww.epfl.ch/pisani/mlib/
23. Pisani MB (2007) Copper/low-K technological platform for the fabrication of high quality factor above-IC passive devices. PhD thesis, EPFL, Lausanne, Switzerland. URL http://library.epfl.ch/theses/?nr=3831
24. Pisani MB, Hibert C, Bouvet D, Dehollain C, Ionescu AM (2005) Fabrication and electrical characterization of high performance copper/polyimide inductors. In: PhD Res Microelectron Electron (PRIME'05), Lausanne, Switzerland, vol 1, pp 185–188
25. Rabjohn GG (1991) Monolithic microwave transformers. M Eng thesis, Carleton University, Ottawa, ON, Canada
26. Spencer RG (2001) Analysis of the modified MOS Wilson current mirror: a pedagogical exercise in signal flow graphs, Mason's gain rule, and driving-point impedance techniques. IEEE Trans Educ 44(4):322–328
27. Sun Y, Fidler JK (1996) Design method for impedance matching networks. IEE Proc Circ Devices Syst 143(4):186–194
28. Yue CP, Wong SS (1998) On-chip spiral inductors with patterned ground shields for Si-based RF IC's. IEEE J Solid-State Circ 33(5):743–752

Chapter 7
Measurement Results for the Frequency-Tunable CMOS RF Power Amplifier

Abstract The experimental characterization and measurement of the frequency-tunable RF power amplifier is the subject of this chapter. The frequency-tunable behavior could not be observed in the characterization of the integrated circuit at 3.7 and 5.2 GHz because of the low coupling factor of the integrated coupled inductors. A hybrid implementation using an integrated stand-alone CMOS power amplifier designed for test purposes, a discrete commercial RF transformer, and a discrete bipolar transistor to control the current in the secondary winding of the transformer was carried out. The circuit was designed for operation at 200 and 300 MHz. The measurements have shown that at 200 MHz a relative improvement in efficiency of a factor of 1.4 was achieved. Moreover, less distortion was generated with the proposed tunable output matching network. The hybrid implementation allowed us to demonstrate the feasibility of the frequency-tunable RF power amplifier based on coupled inductors.

7.1 Introduction

This chapter begins, in Sect. 7.2, with the characterization of a stand-alone PA equal to the oned used as the core of the frequency-tunable amplifier. It continues then with the description of the characterization procedure for the integrated coupled inductors in Sect. 7.3. As seen in the previous chapters, the coupling factor and the value of the inductors are of major importance for the operation of the tunable PA. The measurement of the tunable PA is presented in Sect. 7.4. The results obtained were unsatisfactory. Hence, a hybrid implementation using the integrated stand-alone CMOS PA, discrete coupled inductors, and a control circuit built with a discrete bipolar transistor was carried out to validate the concept. This is the subject of Sect. 7.5.

P.A. Dal Fabbro, M. Kayal, *Linear CMOS RF Power Amplifiers for Wireless Applications,*   101
Analog Circuits and Signal Processing,
DOI 10.1007/978-90-481-9361-5_7, © Springer Science+Business Media B.V. 2010

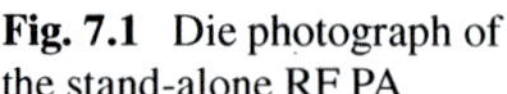

Fig. 7.1 Die photograph of the stand-alone RF PA

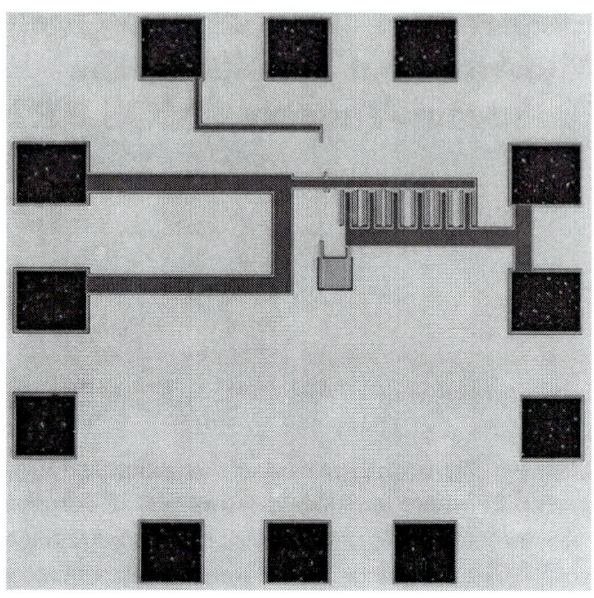

7.2 Stand-Alone RF Power Amplifier Characterization

A stand-alone PA identical to the one used as the core of the frequency tunable RF PA was designed and measured. Its schematic was already shown in Fig. 6.5 in the previous chapter (refer to p. 84) and a die photograph is depicted in Fig. 7.1. The integrated components are the transistor (M_1), the stability resistor (R_{stab}) and capacitor (C_{stab}), and the bias resistor (R_{bias}). The matching networks, the DC blocks, and the RF choke were placed off-chip on the printed-circuit evaluation board.

The PA chip was bonded to the board using the special procedure described in Sect. 7.4. Its input and output were connected to the ports of the network analyzer through SMA connectors, coaxial bias tees, and coaxial cables. The bias tees were used to provide gate (input) and drain (output) bias voltages for transistor M_1 as well as DC blocks for the input and output.

The input and output matching networks were designed for a simultaneous conjugate match at 5.2 GHz using the procedure described in Chap. 10 (Appendix B). The 2-port S parameters were measured with the network analyzer HP8719D [1] calibrated from 1 to 6 GHz with the SOLT calibration kit [2] provided by the manufacturer. The S parameters corresponding to the matched condition are shown in Figs. 7.2 and 7.3. It is worth stressing that this condition corresponds to a conjugate match at 5.2 GHz only, meaning that better matching could be achieved with different matching networks at other frequencies.

Table 7.1 summarizes the S-parameter values for the PA matched for a 5.2 GHz operation. Values at 2.4, 3.7, and 5.2 GHz are shown.

After that the PA was matched at 5.2 GHz, a 2-tone test was performed to measure the efficiency and linearity. Two tones centered at 5.2 GHz and spaced 500 kHz apart from each other were applied. The input power was swept and the measured

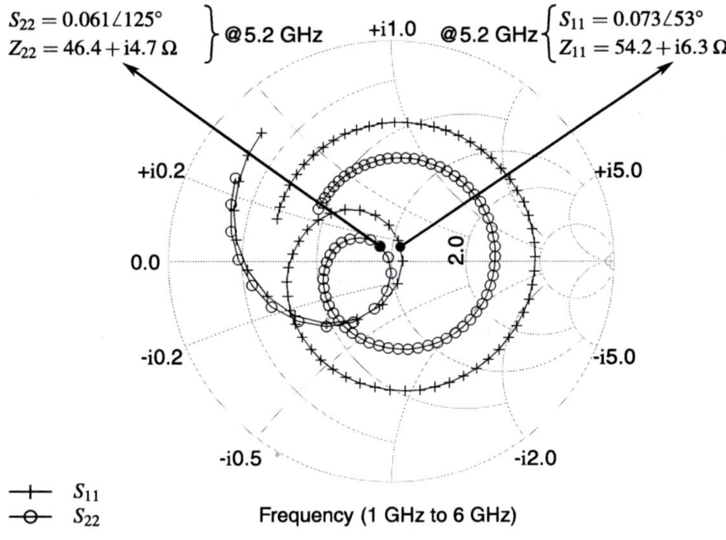

Fig. 7.2 SOLT measurement of the stand-alone PA matched at 5.2 GHz (S_{11} and S_{22})

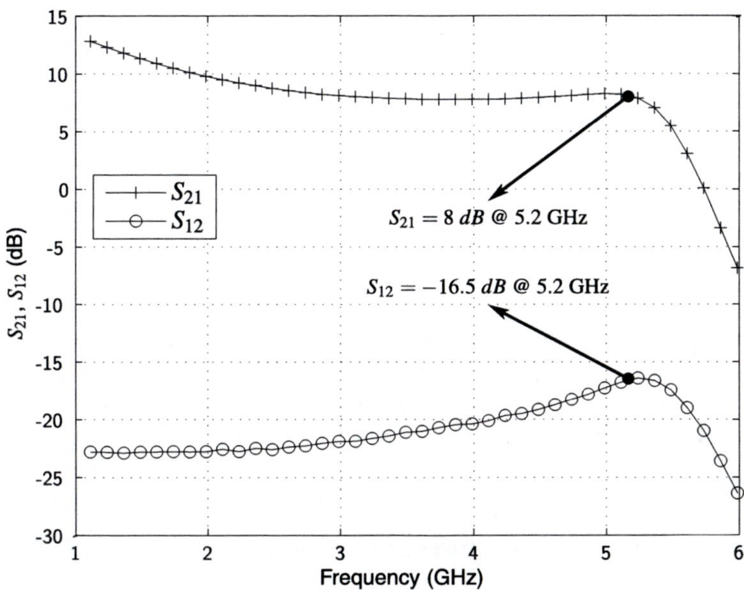

Fig. 7.3 SOLT measurement of the stand-alone PA matched at 5.2 GHz (S_{21} and S_{12})

power gain is shown in Fig. 7.4. The values are in accordance with the small-signal S-parameters measurement. The IMD3 and PAE were measured for each input power and are shown in Fig. 7.5. The −35 dBc linearity limit is also shown (*dashed line*). A summary of these results is given in Table 7.2.

Table 7.1 Measured S parameters for the stand-alone PA. Values for the PA matched at 5.2 GHz

SP	2.4 GHz	3.7 GHz	5.2 GHz
S_{11} (dB)	−3.7	−4.5	−22.8
S_{22} (dB)	−6.3	−7.7	−24.2
S_{21} (dB)	8.9	7.8	8
S_{12} (dB)	−22.5	−20.8	−16.5

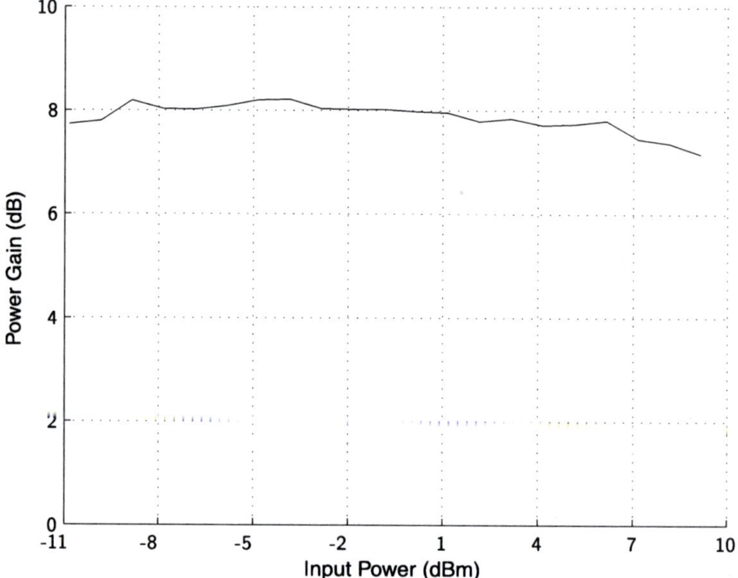

Fig. 7.4 Measured 2-tone power gain at 5.2 GHz for the stand-alone RF PA

We can see that the maximum linear output power delivered by the stand-alone PA in a simultaneous conjugate match condition is about 14 dBm (25 mW). The PAE at this distortion level is 9.7%. These values show that the stand-alone PA operates satisfactorily.

7.3 Coupled Inductors Characterization

A test structure composed of the coupled inductors configured as a 3-terminal device was designed as shown in Fig. 7.6. The purpose was to allow an on-wafer characterization with a probe station. One terminal of each winding was used as an RF port, whereas the second terminal of each winding was grounded. Figure 7.7 shows a photograph of the test structure. A Ground-Signal-Ground (GSG) pad configuration was employed. This procedure of characterization is based on [11] and is suitable for 2-port measurements. For pad de-embedding, an open structure was used.

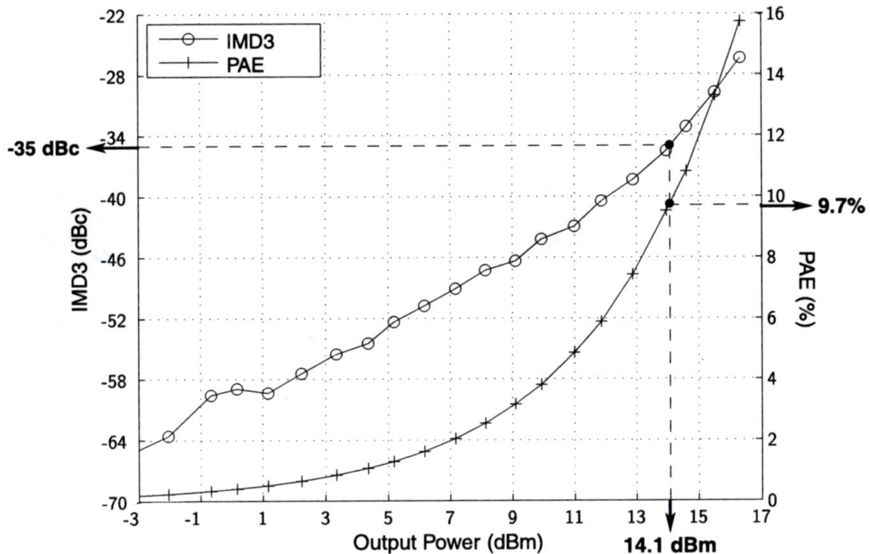

Fig. 7.5 Measured 2-tone IMD3 and PAE at 5.2 GHz for the stand-alone RF PA. IMD3 limit of −35 dBc is also shown (*dashed line*)

Table 7.2 Summary of measured 2-tone results at −35 dBc IMD3 for the stand-alone RF PA operating at 5.2 GHz

Parameter	Value
Input power (dBm)	6.3
Gain (dB)	7.7
Output power (dBm)	14.1
PAE (%)	9.7

Fig. 7.6 Three-terminal transformer configuration for on-wafer characterization

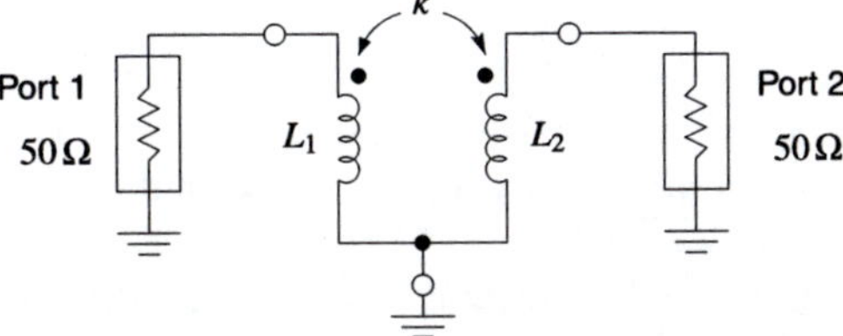

The characterization was performed through S-parameters measurement using the network analyzer HP8719D [1] connected to the SUSS MicroTec cryogenic prober PMC150 [15]. Full 2-port calibration was performed prior to the measurements.

The physically-based equivalent circuit model for such a structure [12, Fig. 22] is given in Fig. 7.8. This circuit can be divided in two parts. The first comprises the components that can be extracted with Low-Frequency (LF) measurements: L_1, R_{Ls1}, L_2, R_{Ls2}, and k. The second consists of the remaining components that are

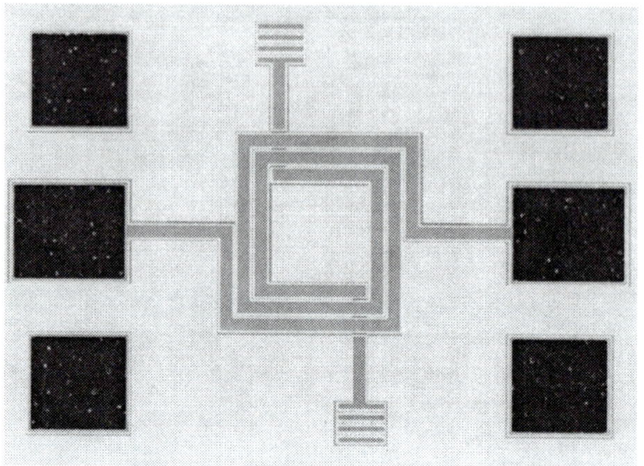

Fig. 7.7 Photograph of the square transformer for on-wafer characterization

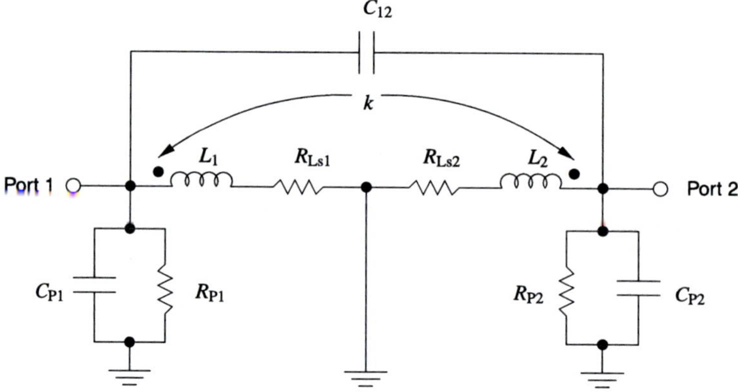

Fig. 7.8 Equivalent circuit for the 3-terminal transformer

extracted with High-Frequency (HF) measurements: C_{12}, C_{P1}, R_{P1}, C_{P2}, and R_{P2}. To further simplify the extraction procedure, a T model [9] is used in the LF part to represent the magnetic coupling between the two inductors as shown in Fig. 7.9— with $M = k\sqrt{L_1 L_2}$.

From the low-frequency S parameters, the components of the T model can be extracted by first transforming the S-parameter matrix into a Z-parameter matrix and then using the equations shown in Fig. 7.10 [13, Chap. 4], [3].

From the extracted LF components, a Y-parameter matrix (Y_{ext}) is created for the whole frequency range used for the measurements to allow the extraction of the HF components shown in Fig. 7.11. For this purpose, the S parameters from the HF measurements are first transformed into Y parameters (Y_{meas}) and, then, the Y-parameter matrix is subtracted from them ($Y_{diff} = Y_{meas} - Y_{ext}$).

The LF and HF measurements for the model extraction were performed at 50 MHz and 10 GHz, respectively. The extracted square transformer parameters corresponding to the model of Fig. 7.8 are shown in Table 7.3.

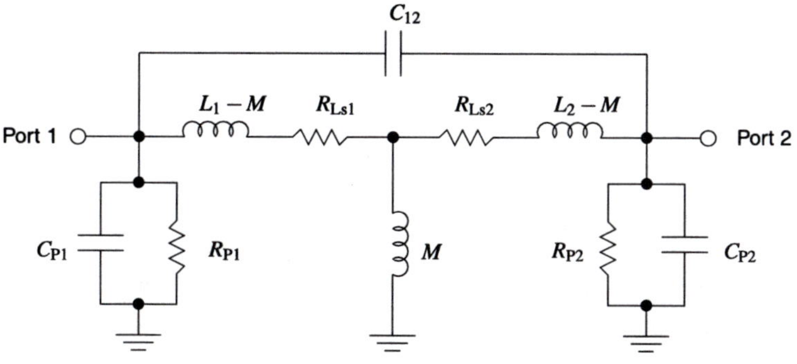

Fig. 7.9 Equivalent T model for the 3-terminal transformer

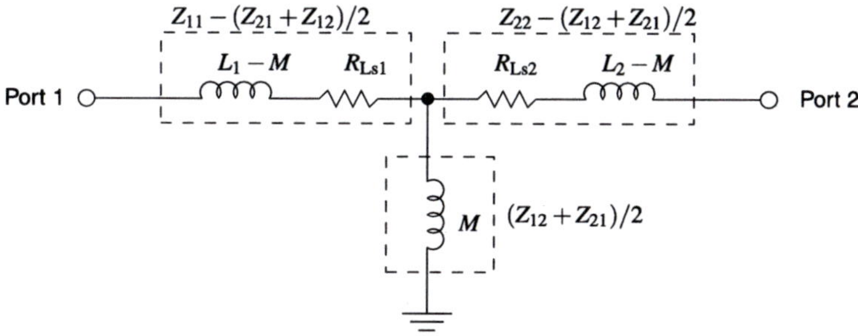

Fig. 7.10 Low-frequency equivalent T model for the 3-terminal transformer parameter extraction

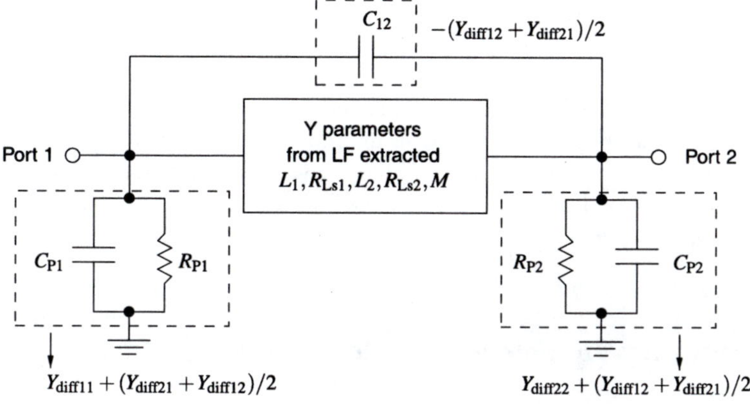

Fig. 7.11 High-frequency equivalent circuit for the 3-terminal transformer parameter extraction

The model of Fig. 7.8, with the parameters of Table 7.3, was simulated and compared with the measurement data. The parameters S_{21} and S_{11} are shown in Figs. 7.12 and 7.13. The measured and extracted characteristics are in good agreement if we consider that the values of the components in the model are constant with

Table 7.3 Extracted model parameters for the interleaved square transformer

Component	Value
L_1	0.69 nH
L_2	0.69 nH
R_{Ls1}	2.1 Ω
R_{Ls2}	2.1 Ω
k	0.38
C_{12}	43 fF
C_{P1}	15 fF
C_{P2}	15 fF
R_{P1}	1.7 kΩ
R_{P2}	1.7 kΩ

frequency. The maximum magnitude and phase errors are 2 dB and $10°$ for S_{21}, and 0.5 dB and $2°$ for S_{11}, from 1 to 10 GHz. The transformer is considered a reciprocal device with $S_{11} = S_{22}$ and $S_{21} = S_{12}$.

Note that we have used an open structure as the only mechanism to de-embed the pad. However, Gan in [6] states that

> the open-only de-embedding procedure generally overestimates the inductance of the windings and underestimates the coupling coefficient k.

He claims that

> the open-short de-embedding procedure results in a much better match between the simulated data and the measured data ....[1]

We see from Table 7.3 that the coupling factor is much lower than the one we used in the design of the frequency-tunable PA—0.38 against 0.7. However, it is in agreement with the simulation results obtained with ADS Momentum, which gives $k = 0.39$. This information, as stated in the previous chapter, was not available at the time of the design. We investigated the model of the transformer after that the measurement results showed that the tunable PA did not work as expected, as it will be seen in the next section.

7.4 Frequency-Tunable RF PA Measurement

The photograph of the frequency-tunable RF PA chip is shown in Fig. 7.14. (Refer back to Fig. 6.17 on p. 96 to identify the component designators used.) The RF Choke (RFC), C_3, and L_3 are off-chip components. R_{stab}, R_{bias}, and R_{bias2} are not

---

[1]In our case, we can say that the open-only pad de-embedding method used was enough because the coupling factor predicted by the 2.5D electromagnetic simulator was very close to the measured value.

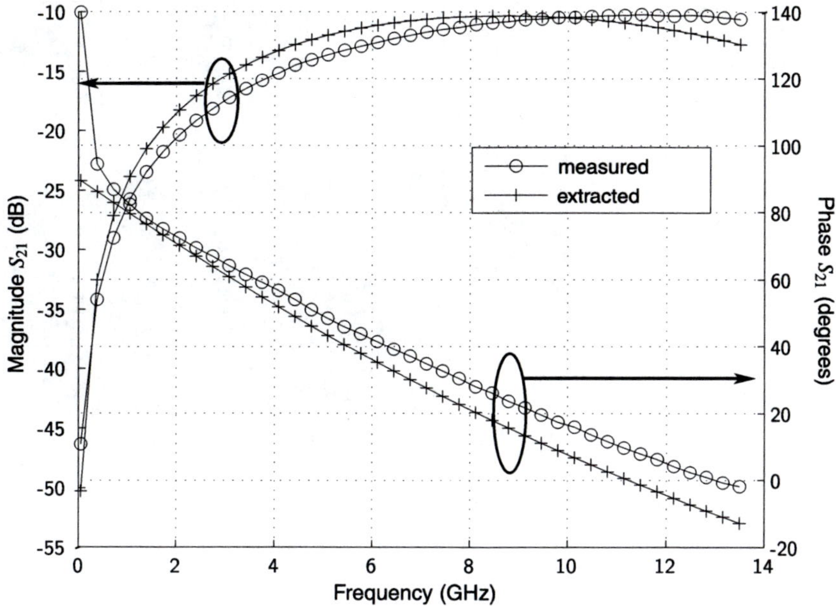

Fig. 7.12 Square transformer characterization results (S_{21})

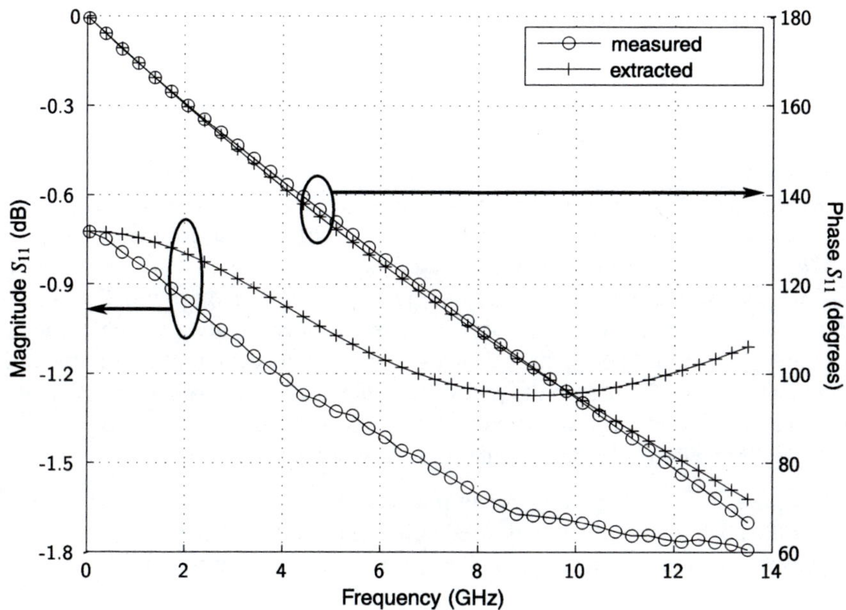

Fig. 7.13 Square transformer characterization results (S_{11})

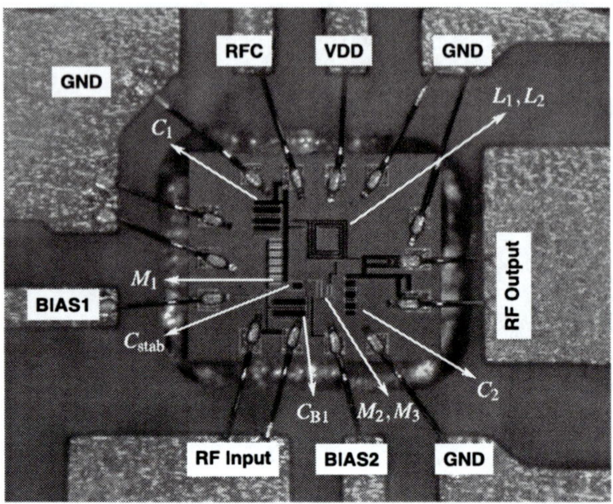

Fig. 7.14 Photograph of the tunable RF PA chip bonded to a printed-circuit board

visible. The integrated circuit occupies an area of 1.2 mm^2 (1200 µm per 1000 µm) including the 14 pads used. Six of them are used for the ground connection. A high number of ground pads is desirable to reduce the inductance between the chip and PCB grounds. Although the VDD supply voltage is not directly connected to the integrated circuit, as it arrives from the off-chip RF Choke, a VDD pad is used to bias the deep n-well[2] with the highest DC voltage in the circuit.

In a similar way to the dynamic supply PA, the chip was lodged in a window opened in the printed-circuit board in order to reduce the length of the bondwires resulting from the chip-on-board mounting. Bondwires with 0.4 µm length could be realized with this method, resulting in inductances of approximately 0.4 nanohenry per bondwire, if we consider a value of 1 nH/mm [8, p. 53]. (A more accurate calculation of bondwire inductance can be done as described in [10].) This special mounting procedure involved opening a window in the board and soldering a piece of metal sheet at the bottom of the board below the window. The chip was inserted in the window with its bottom attached to the metal sheet with a conductive glue. This was done so as to connect the back of the chip to the ground plane of the PCB (bottom layer). The PCB was fabricated using Rogers high-frequency laminate RO4350 [14] whose main properties are given in Table 4.1 on p. 40. This kind of laminate is distinguished for its stable electrical properties and low dielectric loss (low $\tan \delta$ [13, Chap. 1]), which makes it suitable for the design of repeatable high-frequency circuitry. To reduce the parasitic inductance in the ground lines, multiple vias were used to connect the ground in the top layer of the PCB to the ground plane.

The test procedure was done in the following order:

1. Bias the PA at the desired quiescent point (through V_{bias1}).
2. Bias the control circuit at the desired quiescent point (through V_{bias2}).
3. Measure the S parameters and implement the input impedance matching.

---

[2]In triple well processes, the deep n-well is used as an RF isolation strategy for n-type devices [4].

Table 7.4 Quiescent point for the tunable power amplifier

Voltage/Current	Value
V_{bias1}	870 mV
I_{d1}	86 mA
V_{bias2}	650 mV
I_{ctrl_q}	4.7 mA

Table 7.5 Main measured S parameters for the tunable and conventional PAs

SP	2.4 GHz		3.7 GHz		5.2 GHz	
	Conv.	Tun.	Conv.	Tun.	Conv.	Tun.
S_{11} (dB)	−3.5	−3.4	−3.7	−3.7	−11.8	−10.6
S_{22} (dB)	−11.8	−11.3	−11.6	−11.6	−15.7	−16.3
S_{21} (dB)	8.2	7.8	6.4	6	3.6	3.3
S_{12} (dB)	−27.8	−28.7	−26.6	−25.5	−26.3	−26.6

4. Perform 2-tone measurements of the tunable PA.
5. Set V_{bias2} to zero to disable the control circuit.
6. Perform S-parameters measurement to verify the input matching and change it if needed.
7. Perform 2-tone measurements of the conventional PA.

Table 7.4 shows the quiescent currents and voltages for the PA and the control circuit.

The control voltage V_{bias2} changes the transconductance of the control circuit, thereby altering the relationship between the currents I_{ctrl} and I_{RF} according to (6.27) on p. 92. Hence, by varying V_{bias2} it was possible to find a quiescent point corresponding to an optimum performance for the tunable PA at each frequency of operation.

The 2-port S parameters were measured with the network analyzer HP8719D [1] calibrated from 1 to 6 GHz with the SOLT calibration kit [2] provided by the manufacturer. The input (S_{11}) and output (S_{22}) reflection coefficients are shown in the Smith charts of Figs. 7.15 and 7.16. The magnitudes of the measured forward (S_{21}) and reverse (S_{12}) power gains are shown in Fig. 7.17. These three figures reveal that there is almost no difference between the behavior of the tunable and conventional PAs. This indicates that there is a problem with the frequency-tunable PA that must be investigated. A summary of the results is shown in Table 7.5.

Figure 7.18 shows the 2-tone measurement results for the frequency tunable PA. The frequency was swept from 1 to 5 GHz in 0.5 GHz steps. The output power and PAE were measured at an IMD3 level of −35 dBc. The performance of the tunable PAE is better at every frequency. However, a frequency-tunable behavior—as if the PA could be tuned for optimum operation at different frequencies—is not observed.

Figure 7.19 shows the same measurement, but for a narrow range of frequencies. In this figure, the frequency was swept from 2.2 to 2.6 GHz in 50 MHz steps. For

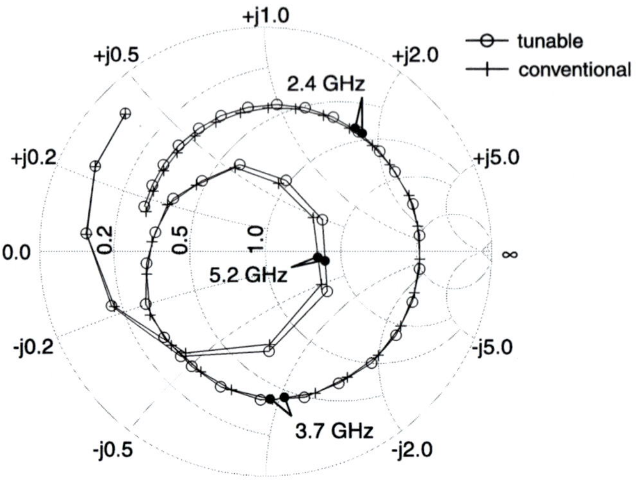

Fig. 7.15 Measurement of S_{11} for the tunable and conventional PAs

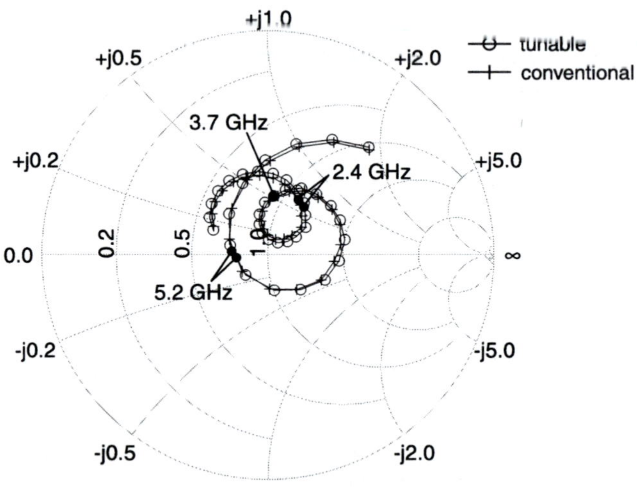

Fig. 7.16 Measurement of S_{22} for the tunable and conventional PAs

the tunable PA, both the output power and PAE at -35 dBc IMD3 are improved by a fixed amount with respect to the conventional one. Here again no frequency-tunable behavior could be observed.

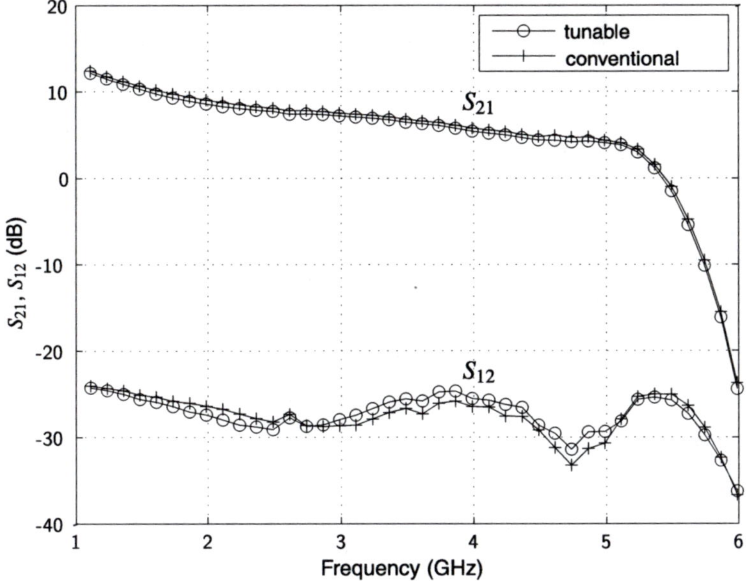

Fig. 7.17 Measurement of S_{21} and S_{12} for the tunable and conventional PAs

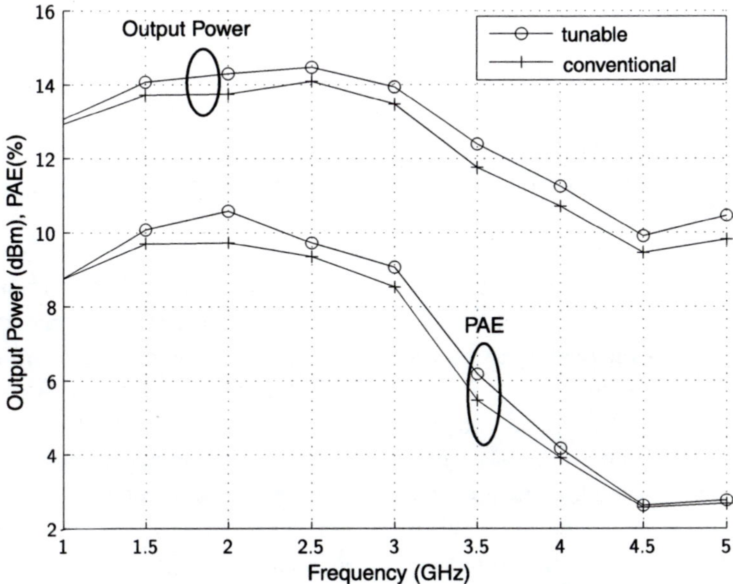

Fig. 7.18 Measurement results for the tunable PA for a wide frequency sweep

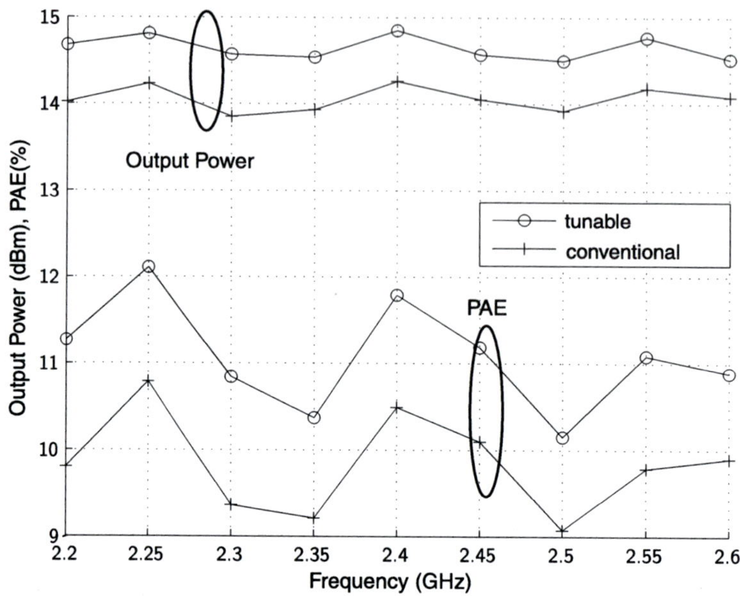

Fig. 7.19 Measurement results for the tunable PA for a narrow frequency sweep

7.5 Hybrid Implementation

The circuit of a hybrid implementation of the tunable RF PA is shown in Fig. 7.20. It was designed for the 200 and 300 MHz frequency bands—tuning range of 40%. We used the stand-alone CMOS PA described in Sect. 7.2 biased in class AB with a quiescent drain current of 25 mA. The input and output matching networks, the RF choke, and the control circuit were implemented with discrete components.

Inductors L_1 and L_2 were implemented with the 1812WBT-5L RF transformer [5], having a winding inductance of 90 nH, a coupling factor of unity, and a series parasitic resistance of 0.15 Ω. The high values of winding inductances for the available commercial transformers limited the frequency of operation of the circuit and the minimum value of R_{opt}. The bipolar transistor BFP405 [7] was used to implement the control circuit with a quiescent collector current of 1.5 mA.

The target output power was set to 10 dBm and an optimum resistance of 150 Ω was calculated using the loadline method. Load-pull simulations were also used to determine the values that the load impedance could take. $C_1 = 12$ pF and $C_2 = 18$ pF provide the desired load impedance when 50 nH $< L_{eff} < 67$ nH at 200 MHz and 29 nH $< L_{eff} < 36$ nH at 300 MHz. L_{adj} is used to set the value of L_{eff}, and equals 220 nH at 200 MHz and 51 nH at 300 MHz.

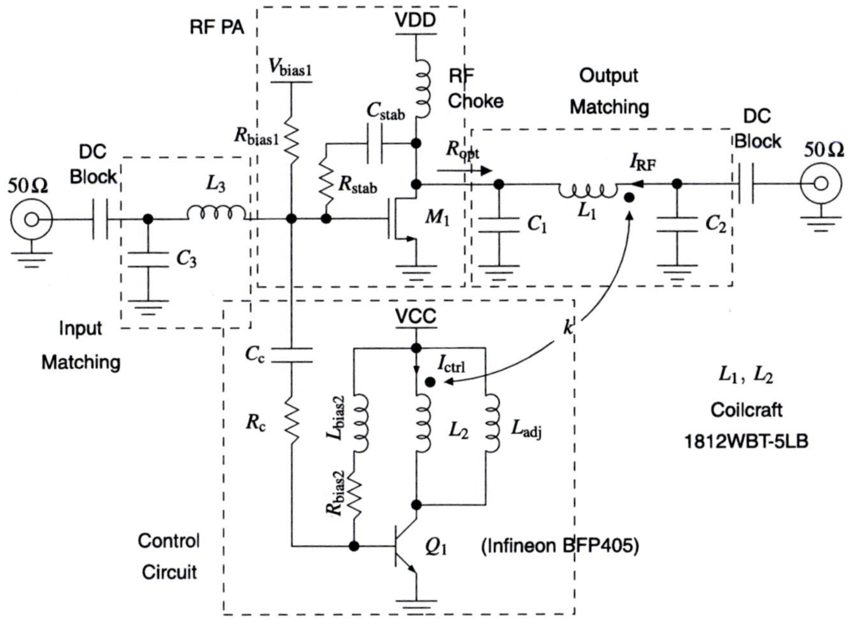

Fig. 7.20 Hybrid frequency-tunable RF power amplifier schematic

7.5.1 Measurement Results for the Hybrid PA

In order to evaluate the benefits of the tunable PA, the conventional power amplifier was also measured. It uses the same PA with the control circuit disabled. The transformer is replaced by a chip inductor of the value required to attain $R_{opt} = 150\ \Omega$ at 300 MHz. Figures 7.21 and 7.23 show the IMD3 results for the 2-tone measurements with tones centered at 200 and 300 MHz, and 500 kHz tone spacing. The corresponding PAE results are given in Figs. 7.22 and 7.24.

At 300 MHz, although the conventional PA was expected to present a better performance, the tunable PA generates less distortion with almost equal PAE (at the same output power) for output power levels greater than 7 dBm (5 mW). At 200 MHz, the tunable PA shows much lower IMD3 than its conventional counterpart for output power levels lower than 11.2 dBm (13.2 mW). These measurement results are summarized in Table 7.6.

At an IMD3 level of -30 dBc, the tunable PA attains 10.5 dBm (11.2 mW) output power and 15% PAE at 200 MHz, whereas the conventional PA attains 5.9 dBm (3.9 mW) and 4.9%. The ratio between the efficiencies of the tunable and conventional PAs is also shown in Fig. 7.22, revealing that at 200 MHz, a relative improvement of a factor of 1.38 (PAE ratio = 138%) in PAE is attained using the tunable PA.

At 300 MHz and -30 dBc IMD3, the tunable PA presents an output power of 10.1 dBm (10.2 mW) and a PAE of 15% against 7.8 dBm (6 mW) and 8.5% of the conventional PA. The ratio between the efficiencies of the tunable and conventional

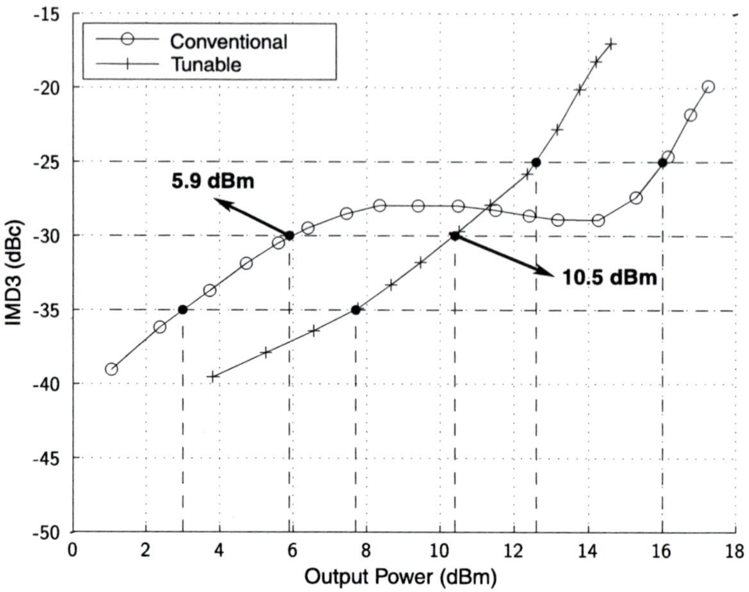

Fig. 7.21 Measured IMD3 at 200 MHz. IMD3 levels of −25, −30, and −35 dBc are highlighted (*dashed lines*)

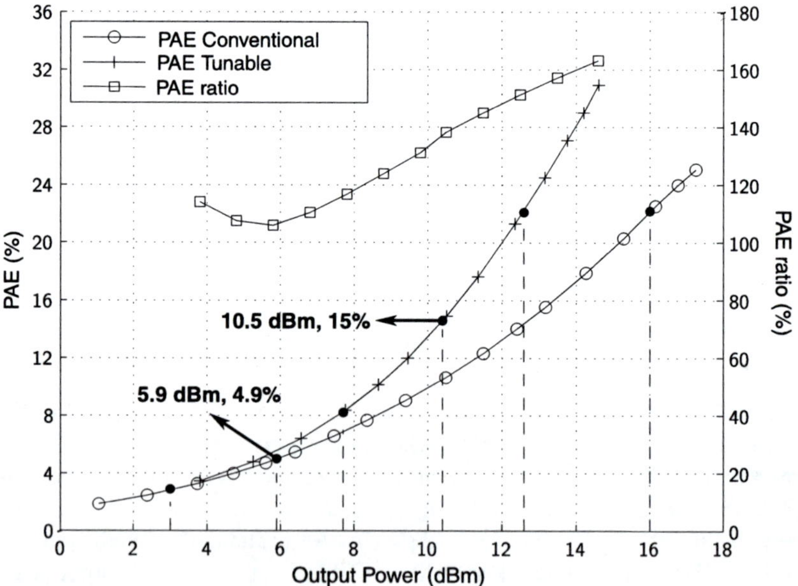

Fig. 7.22 Measured PAE at 200 MHz

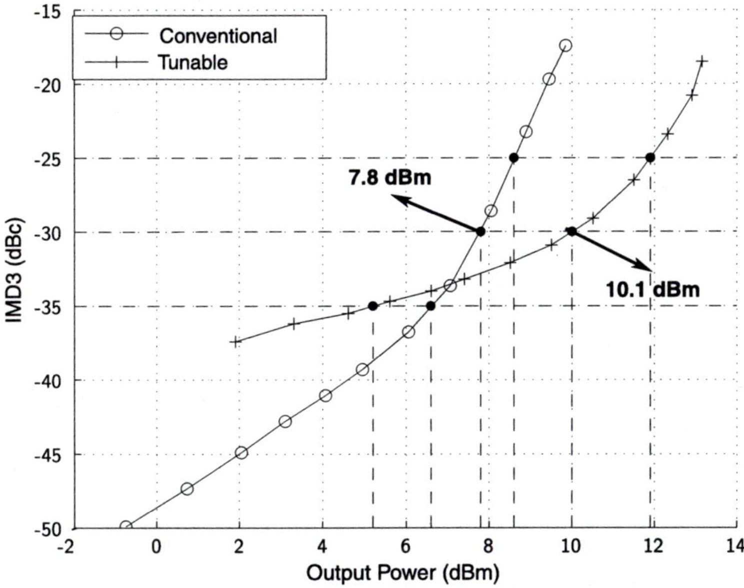

Fig. 7.23 Measured IMD3 at 300 MHz. IMD3 levels of −25, −30, and −35 dBc are highlighted (*dashed lines*)

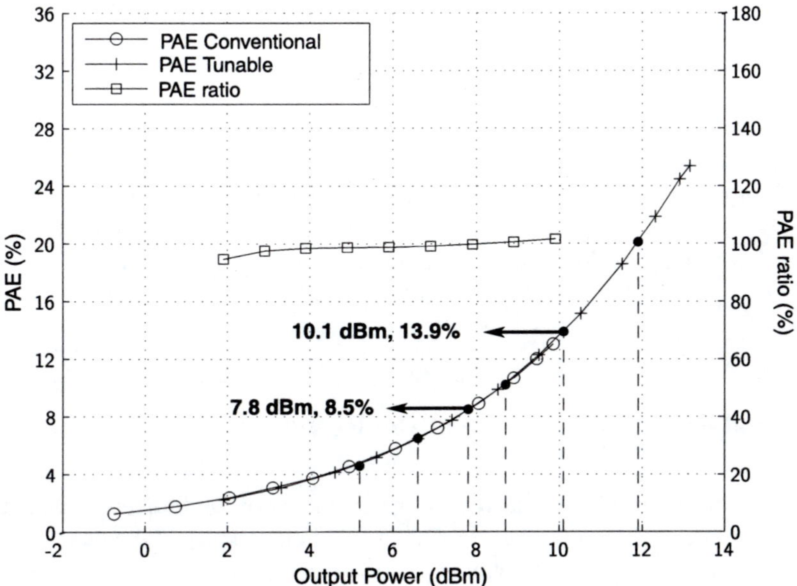

Fig. 7.24 Measured PAE at 300 MHz

Table 7.6 Measured 2-tone output power and PAE

2 Tone	200 MHz		300 MHz		IMD3
	Conv.	Tun.	Conv.	Tun.	value
P_{out} (dBm)	3.1	7.6	6.6	5.1	
PAE (%)	2.8	7.9	6.5	4.6	−35 dBc
P_{out} (dBm)	5.9	10.5	7.8	10.1	
PAE (%)	4.9	15	8.5	13.9	−30 dBc
P_{out} (dBm)	16	12.5	8.7	11.9	
PAE (%)	22.1	21.9	10.2	20.1	−25 dBc

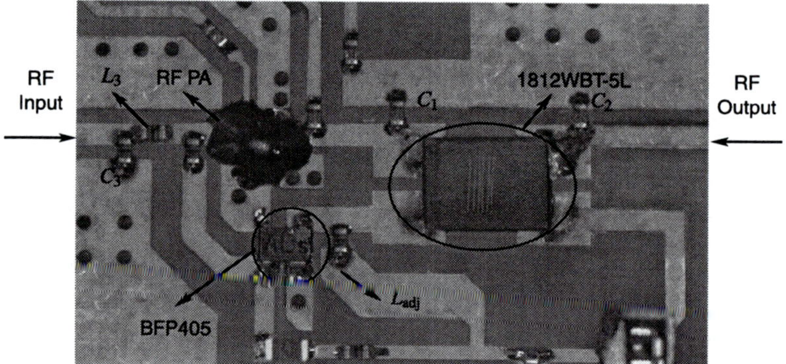

Fig. 7.25 Photograph of the PCB used in the hybrid implementation

PAs, shown in Fig. 7.24, reveals that at 300 MHz, there is almost no difference between their efficiencies.

A photograph of the printed-circuit board (PCB) for this hybrid implementation is shown in Fig. 7.25. Due to the lower frequencies at which the circuit was tested, an FR-4 substrate (thickness of 0.5 μm, copper cladding of 17 μm, and ϵ_r of 4.6) was chosen.

7.6 Analysis and Discussion of the Results

The experimental results obtained with the prototype of the frequency-tunable RF power amplifier presented in Sect. 7.4 showed that the expected frequency-tunable behavior did not take place. Looking back at the design, we can say that there were two main problems that we overlooked and which are intimately related to each other.

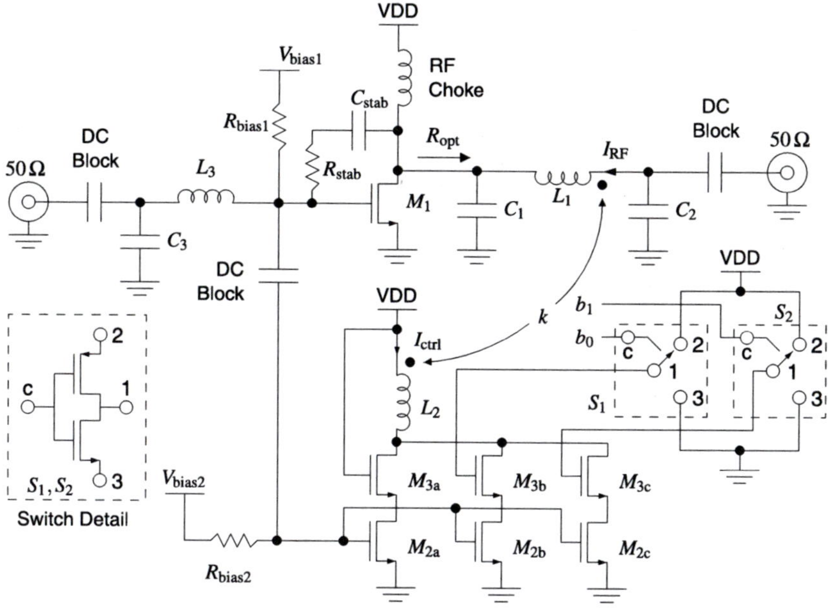

Fig. 7.26 Suggested frequency-tunable RF PA with adjustable control circuit

> **Layout Parasitics** Although we knew that layout is of fundamental importance in RF IC design, its consequences were devastating in our case because slight variations on parasitic capacitances at the output matching network can play havoc with the desired transformed impedance.
>
> **Coupled Inductors** The design of coupled inductors is a difficult subject and if the achieved coupling factor is different from the predicted value, the transformed impedance can be very different from the desired value.

These two issues directly affect the transformed impedance and, hence, the performance of the amplifier. Therefore, it is a hard task to design a frequency-tunable PA based on coupled inductors and achieve a good performance in the first design.

Moreover, the only parameter that we had control to check the circuit performance was the control circuit bias voltage. Although this voltage affects the transconductance g_{m2} of the control circuit according to (6.27) on p. 92, it would have been safer to foresee another possibility of adjustment in case of malfunctioning. Our suggestion is to add to the control circuit parallel branches that can be switched on and off, as indicated in Fig. 7.26, thereby changing the current I_{ctrl}. Furthermore, the capacitances associated to these extra branches change the equivalent impedance of node v_y in Fig. 6.12 on p. 90. Changing the impedance of this node alters the current ratio responsible for the final value of the tunable inductance.

With these considerations, a possible strategy is to make a first design in which the integrated transformer is carefully simulated with an electromagnetic software—

which showed results very close to the experimental values in our case. In this first design, both the PA and the transformer should be integrated. This first integration allows the characterization of the RF power amplifier and the coupled inductors separately. Then, in a second design, the complete system would be integrated following the schematic of Fig. 7.26.

Although the fully-integrated version of the frequency-tunable PA did not work, the hybrid implementation showed that the principle underlying the technique is valid and can be used in tunable power amplifiers. When considering using the technique, the designer should however take into account the important points listed below.

- Required output power: it determines the choice of R_{opt} and, hence, of the matching network components.
- Required frequency bands: it affects the inductance tuning range and the values of C_1, L_1 and C_2. The higher the frequency, the lower the inductor value for fixed C_1 and C_2.
- Required optimum resistance: it depends on the required output power. It affects C_1, L_1 and C_2. For fixed values of C_1 and C_2, the higher is R_{opt}, the higher is the value of L_1 and the lower is the attainable output power for a given VDD.
- Value of the integrated inductor (L_1): it must be within a range that can be integrated. Hence, it is limited by the required R_{opt} and by IC technology constraints.
- Values of C_1 and C_2: for a given R_{opt}, the higher are C_1 and C_2, the higher is Q_0 and, hence, the lower is the difference between the achieved R_{opt} at different frequencies when keeping C_1 and C_2 constant and varying just L_1. However, the higher Q_0 is, the lower is L_1—in this case, L_1 can attain values that are impossible to be implemented.
- Required inductance tuning range: it affects the required α and β variation. The higher α and β variation is, the more difficult is the implementation of the control circuit. The higher is α, the higher is the power consumption of the control circuit.
- Coupling factor: it affects the attainable inductance tuning range. Layout optimization must be done to increase the coupling factor.

7.7 Conclusion

This chapter presented the measurement results for the frequency-tunable RF power amplifier including individual results for each functional block (RF PA and coupled inductors) as well as the results for the complete tunable PA. Since the complete integrated system did not work as expected, a hybrid circuit using the stand-alone RF PA together with a discrete transformer and a discrete bipolar RF transistor to implement the control system was realized. The simulation results of the integrated tunable PA at 3.7 and 5.2 GHz presented in Chap. 6, and the measurement results of the hybrid implementation at 200 and 300 MHz discussed in this chapter, showed that a frequency-tunable impedance matching network based on coupled inductors can be employed in the output of a CMOS RF power amplifier. With such a network the PA can operate in two different frequency bands with superior performance without the need of modifying the matching networks or of using multiple PAs.

References

1. Agilent (1999) 8719D Network Analyzers. Agilent Technologies, USA. URL http://cp. literature.agilent.com/litweb/pdf/08720-90288.pdf
2. Agilent (2007) 85052D 3.5 mm economy calibration kit. User's and Service Guide. URL http://cp.literature.agilent.com/litweb/pdf/85052-90079.pdf
3. Agilent (2010) IC-CAP modeling handbook. URL http://edocs.soco.agilent.com/ display/iccapmhb/Home
4. Blalack T, Leclercq Y, Yue CP (2002) On-chip RF isolation techniques. In: Proc IEEE Bipolar/BiCMOS Circuits Technol Meet (BCTM'02), Monterey, CA, pp 205–211
5. Coilcraft (2006) Surface mount wideband RF transformers. Data Sheet 117-1. URL http://www.coilcraft.com/pdfs/smwbt.pdf
6. Gan H (2006) On-chip transformer modeling, characterization, and applications in power and low noise amplifiers. PhD thesis, Stanford University, Stanford, CA
7. Infineon (2007) BFP405 NPN silicon RF transistor. Data Sheet. URL http://www.infineon. com
8. Lee TH (1998) The Design of CMOS Radio-Frequency Integrated Circuits. Cambridge University Press, Cambridge
9. Long JR (2000) Monolithic transformers for silicon RF IC design. IEEE J Solid-State Circ 35(9):1368–1382
10. Long JR (2004) Packaging RF circuits. Lecture Notes for Advanced Engineering Course on RF IC Design, Lausanne, Switzerland
11. Mohan SS (1999) The design, modeling and optimization of on-chip inductor and transformer circuits. PhD thesis, Stanford University, Stanford. URL http://smirc.stanford. edu/papers/Thesis-mohan.pdf
12. Niknejad AM, Meyer RG (1998) Analysis, design, and optimization of spiral inductors and transformers for Si RF ICs. IEEE J Solid-State Circ 33(10):1470–1481
13. Pozar DM (1998) Microwave Engineering, 2nd edn. Wiley, New York
14. Rogers (2006) RO4000 Series High Frequency Circuit Materials. Data Sheet 92-004. URL http://www.rogerscorp.com/acm/literature.aspx
15. SUSS (2003) PMC150–Manual cryo prober. Data Sheet. URL http://www.suss.com

Chapter 8
Conclusion

Abstract This chapter concludes the book by highlighting the results obtained with the two different designs involving a linear CMOS RF power amplifier. For the dynamic supply RF power amplifier, subject of Chaps. 2 to 4, the measurement results showed a relative improvement in efficiency of a factor of 1.7 at 5.2 GHz and of 2.2 at 2.4 GHz, when compared with the same amplifier operating with a constant 2.5 V supply voltage. For the frequency-tunable RF power amplifier, subject of Chaps. 5 to 7, simulations results for the complete CMOS solution operating at 3.7 and 5.2 GHz and the experimental data obtained for an hybrid (discrete + CMOS) prototype operating at 200 and 300 MHz, led to the conclusion that a frequency-tunable RF PA based on coupled inductors is a good solution for a multiband PA. The main contributions of this book and ideas for future works on the subjects that it covers are also highlighted.

8.1 Highlights

This book treated two important aspects of the design of CMOS RF power amplifiers: efficiency enhancement and frequency-tunable capability.

Chapters 2 to 4 presented the design, implementation, and experimental characterization of a dynamic supply CMOS RF PA. The importance of finding a solution for the linearity–efficiency trade-off in RF PA design was highlighted. Among the different efficiency-enhancement techniques, the dynamic supply PA has an advantage with respect to the ripple requirement for the switched-mode modulator used to vary the supply voltage of the PA.

It was demonstrated through measurement results that a dynamic supply RF PA can be designed using a standard deep-submicron CMOS technology in order to mitigate the linearity–efficiency trade-off inherent to RF PAs. Among the measured characteristics, we highlight the relative improvement in efficiency of a factor of 1.7 at 5.2 GHz and of 2.2 at 2.4 GHz—for low output power levels—achieved with the dynamic supply with respect to the same amplifier operating with a constant supply voltage. The efficiency improvement was also observed when amplifying OFDM

P.A. Dal Fabbro, M. Kayal, *Linear CMOS RF Power Amplifiers for Wireless Applications,*   123
Analog Circuits and Signal Processing,
DOI 10.1007/978-90-481-9361-5_8, © Springer Science+Business Media B.V. 2010

signals at 2.4 GHz. The results confirmed the effectiveness of the dynamic supply technique for CMOS RF PAs.

Chapters 5 to 7 focused on the development of a frequency-tunable RF power amplifier. The definition of such a system was presented as being a PA whose operation is adaptive with respect to the frequency of operation. Normally, such a PA would require different impedance matching networks to be implemented for the different frequencies of operation. The proposed system uses coupled inductors as the means for obtaining this tunable operation and obviating the need of different matching networks.

The coupled inductors work as a variable inductor that is placed in the output impedance matching network of the power amplifier. A control circuit is used to change the phase and amplitude of the current that flows through the secondary winding of the coupled inductors, thereby changing the inductance seen by the circuit connected to the primary winding. The variation of the inductance allows one to maintain the transformation factor of the matching network when the frequency is changed.

A CMOS prototype was fabricated in a 0.11 µm technology. The measurement results did not allow us to prove that the principle works. Hence, we implemented a hybrid system composed of an integrated CMOS power amplifier (designed for testing purposes), a discrete transformer, and a discrete RF bipolar transistor. With this hybrid implementation, we showed that a frequency-tunable RF PA can be built using coupled inductors in its output matching network. However, the frequency of operation had to be decreased to 200–300 MHz (TR = 40%) from the targeted 3.7–5.2 GHz (TR = 34%). The reason for reducing the frequency is that the commercially available transformers were not able to operate at higher frequencies. Furthermore, the high inductance values of these discrete transformers forced the reduction of the maximum output power.

The measurement results from the hybrid system, the simulation results for the complete CMOS solution, and the design methodology proposed form a set of information from which we can conclude that a frequency-tunable RF PA based on coupled inductors is a good solution for a multiband PA.

8.2 Main Contributions

8.2.1 Efficiency Enhancement

The effectiveness of the dynamic supply technique had already been demonstrated in a system using an integrated CMOS modulator and a discrete bipolar RF power amplifier operating at 1.9 GHz for CDMA applications [3]. Hence, the main contribution of Chaps. 2 to 4 of this book is to show that the dynamic supply technique can also be successfully applied in a system where both the modulator and the RF power amplifier are monolithically integrated in a deep-submicron CMOS technology. Furthermore, the system was designed for a higher frequency band, 5.2 GHz, for which the performance metric was the IMD3 for a 2-tone test. It was also shown that the dynamic supply PA can be envisaged for application in WLAN. For this purpose, the EVM obtained from OFDM measurements at 2.4 GHz was used.

8.2.2 Frequency-Tunable Capability

The main contribution of Chaps. 5 to 7 of this book is the demonstration of the use of coupled inductors in a tunable matching network for RF power amplifiers. In the peripheral work that sustains this research, some other minor contributions can also be found. In particular, we demonstrated empirically the property of π-matching networks allowing the variation of only the inductance to keep the transformation factor constant at different frequencies. A comparative study of the available techniques that could be used in the implementation of the tunable RF power amplifier was also carried out.

8.2.3 Impedance Matching

Besides the contributions on efficiency enhancement and frequency tunability, in Chap. 10 (Appendix B) a novel procedure for impedance matching for application in the characterization of printed-circuit RF amplifiers is presented. This procedure was developed during the characterization of the power amplifiers described in this book.

8.3 Future Work

8.3.1 Efficiency Enhancement

Chapter 4 showed that the system that we implemented has some flaws that should be solved for a practical implementation. A first work that should be done is designing the system with the objective of employing it in an integrated CMOS transmitter and not as a stand-alone RF power amplifier. The benefit of doing this is twofold: first, it is a step toward a possible commercial product; second, it obviates the need of analog envelope detection and processing, which would facilitate the characterization of the dynamic supply CMOS RF PA and reveal its true limitations concerning efficiency and linearity for a target application.

A second work that could be done concerns the application in WLAN and consists of designing a higher-order filter to better attenuate the switching signal at the output of the modulator while passing the high bandwidth OFDM envelope signal. The feasibility of such a filter should be studied to determine its specifications for a cutoff frequency of 20 MHz with a switching signal compatible with the global delay of the modulator. It seems that improving the modulator delay from its current

4.5 ns (refer to Sect. 4.3 on p. 42) or 4 ns (Schlumpf's work [3]) would imply in higher power consumption that could cancel out the efficiency improvement of the overall system.

The following quotation, taken from an article that appeared in an electronics magazine [2], evokes the importance of the linearity and efficiency performances in RF PA design for modern communication systems:

> The design of RF power amplifiers is becoming increasingly difficult due to the design requirements for new cellphone modulation schemes .... The demands on linearity and efficiency in these new designs only further accentuate the need for expertise when creating designs that work properly.

8.3.2 Frequency-Tunable Capability

The fact that the prototype of the complete CMOS solution for the frequency-tunable RF PA did not work as expected shows that more work is required to develop a robust system. The main aspects that should be looked at carefully are the integrated transformer design and the layout parasitics. By foreseeing two integrations, the design reliability could be improved. In the first integration, the parasitics of the integrated circuit including pads, bondwires, interconnection capacitance, resistance and inductance, and the transformer design (self-inductance, quality factor and coupling factor) could be evaluated. In the second integration, the complete system could be designed based on the evaluation of the first integration.

To add more flexibility to the system, the two capacitors of the output impedance network could be replaced by variable capacitors (varactors). This would allow a possible error on the value of the tunable inductor to be compensated by changing the value of the capacitors. However, this would increase the system complexity. Nevertheless, these variable capacitors could be used exclusively for calibration purposes.

The following quotation from Huang et al. in a recent paper [1] shows that the quest for tunable RF power amplifiers is still a topical issue:

> Next-generation wireless systems, such as multimode transceivers and cognitive radios, require circuit techniques that facilitate RF adaptivity. Some examples of adaptive circuits include tunable filters, and tunable matching networks for low-noise and power amplifiers.

References

1. Huang C, de Vreede LCN, Sarubbi F, Popadic M, Buisman K, Qureshi J, Marchetti M, Akhnoukh A, Scholtes TLM, Larson LE, Nanver LK (2008) Enabling low-distortion varactors for adaptive transmitters. IEEE Trans Microw Theory Tech 56(5):1149–1163
2. Rako P (2008) Heads and tails: Design RF amplifiers for linearity and efficiency. EDN 53(7):31–38
3. Schlumpf N (2004) Adaptation dynamique de la compression d'un amplificateur RF pour des signaux modulés en amplitude et en phase. PhD thesis, EPFL, Lausanne, Switzerland. URL http://library.epfl.ch/theses/?nr=3020

Chapter 9
Appendix A: Measurement Setups and Additional Screenshots

Abstract This first of two appendices presents the setups used for the measurements performed to characterize the RF power amplifiers treated in this book. The setups for S-parameter, 2-tone, and OFDM measurements are described. It also provides a few additional screenshots taken from the measurement instruments.

9.1 Setups Used for the S-Parameter Measurements

Figure 9.1 depicts one of the setups used for S-parameters measurement. This was the setup used in Sect. 7.2 on p. 102 where no RF choke is employed. In this case, the output bias tee plays the role of both the output DC blocking capacitor and the RF choke inductor. At the input, a bias tee is also used for a similar purpose.

The signal from port 1 of the network analyzer arrives at the evaluation board through an SMA cable (50 cm long and with tight tolerances) and the input bias tee. The output signal arrives from the evaluation board to port 2 of the network analyzer also through an SMA cable and a bias tee. The current flowing through the PA is measured with a true-rms handheld digital multimeter. Three DC voltage sources are used for gate bias, drain bias, and VDD (in fact the triple-well technology requires that the deep n-well be always connected to the highest positive voltage present in the circuit). Three bench-top digital multimeters are used to monitor these voltages.

Figure 9.2 depicts the second setup used for S-parameters measurement. This setup was used in Sect. 4.4 on p. 45 where an RF choke is used for the RF PA drain bias. The difference with respect to the previous setup is that one DC voltage source is eliminated because the drain bias voltage comes also from VDD. The output bias tee is used only for DC blocking with its DC input left unconnected. Table 9.1 lists the instruments used in the S-parameter measurement setups.

9.2 Setup Used for the 2-Tone Measurements

The setup for the 2-tone measurements used to obtain the results for the dynamic supply RF power amplifier presented in Sect. 4.5.1 on p. 50 is depicted in Fig. 9.3.

P.A. Dal Fabbro, M. Kayal, *Linear CMOS RF Power Amplifiers for Wireless Applications,* 127
Analog Circuits and Signal Processing,
DOI 10.1007/978-90-481-9361-5_9, © Springer Science+Business Media B.V. 2010

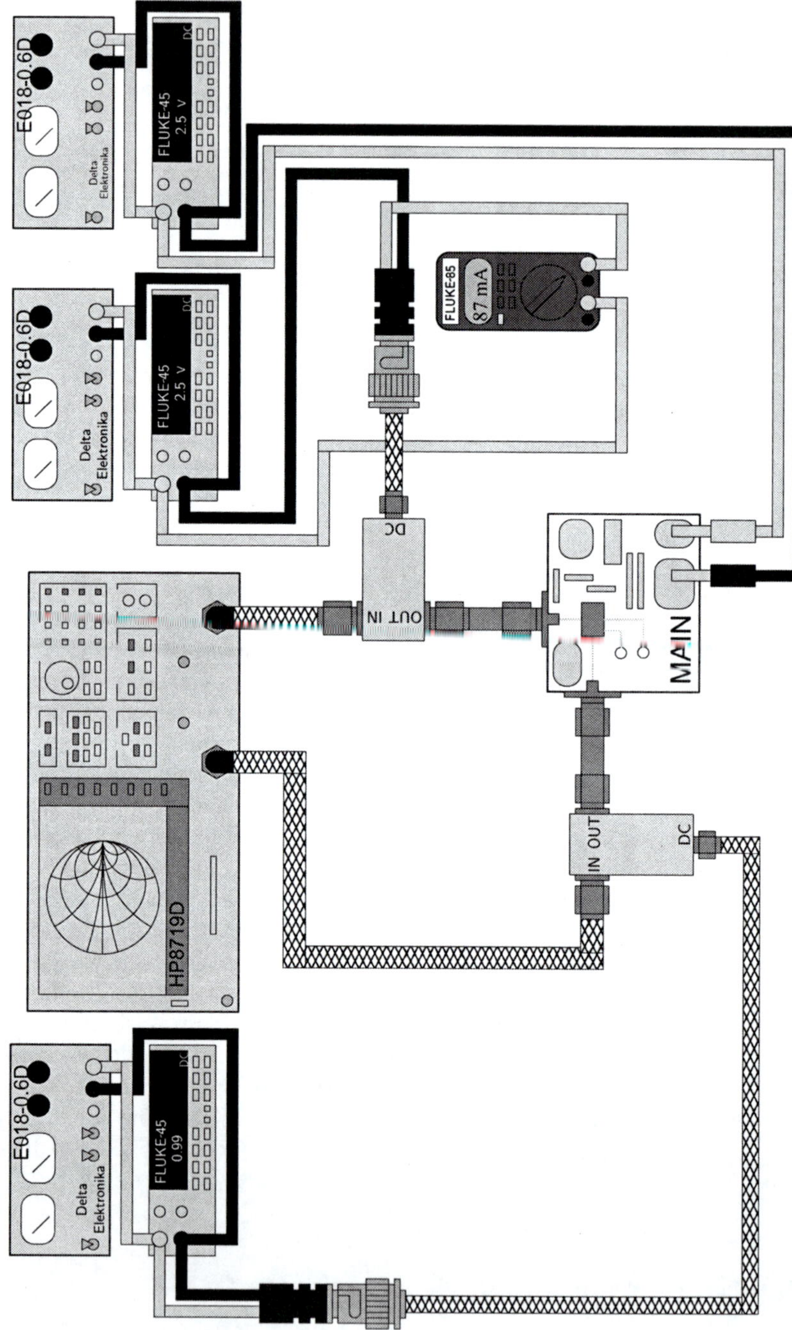

Fig. 9.1 Setup for S-parameters measurement without RF choke

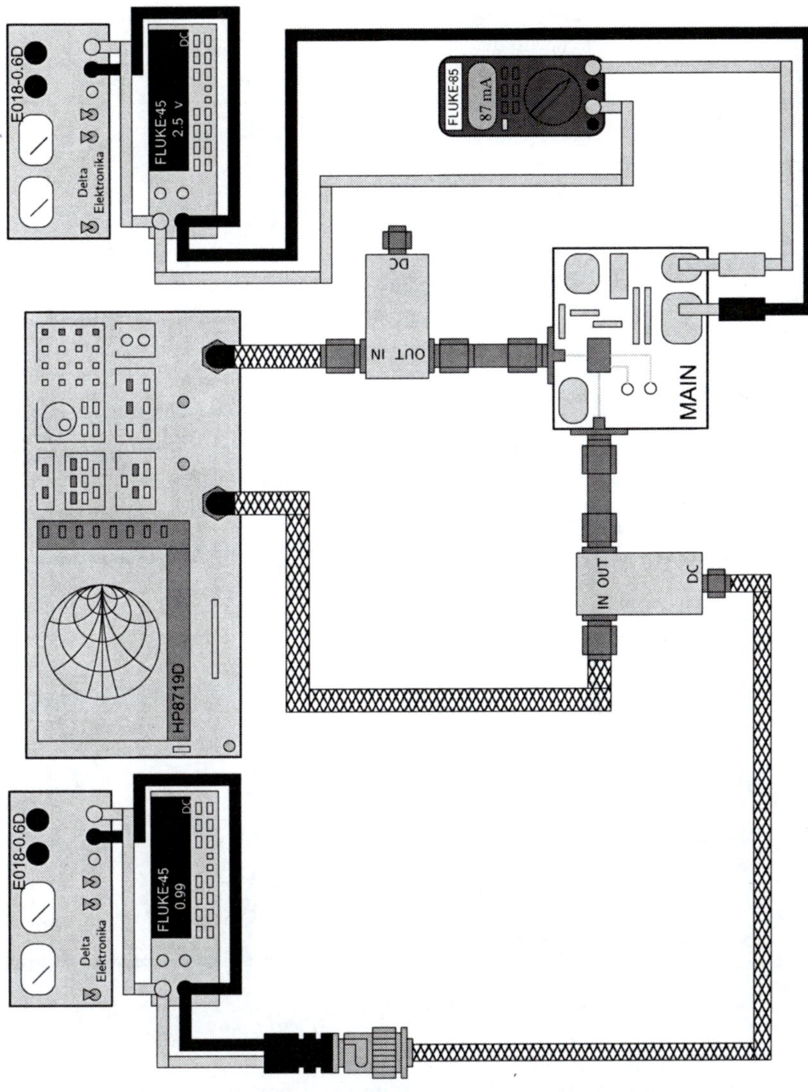

Fig. 9.2 Setup for S-parameters measurement with RF choke

Table 9.1 List of material used in the S-parameter measurement setups

Material	Description
HP8719D [3]	Network analyzer
E018-06.D [8]	DC voltage source
Fluke 45 [9]	Bench-top digital multimeter
Fluke 85 [10]	True-rms handheld digital multimeter
K251 [6]	Ultra-wideband bias-tee
SUCOFLEX 104 [11]	Coaxial cable (SMA)

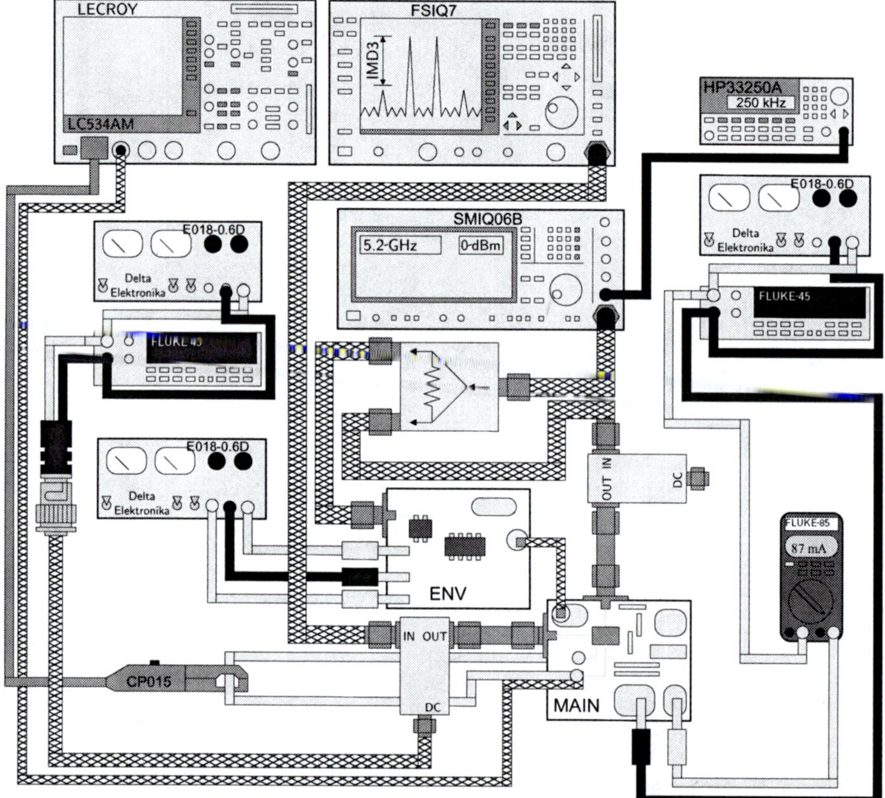

Fig. 9.3 Setup for 2-tone measurements

In this setup, two evaluation boards are used, one that contains the integrated circuit with the modulator and the RF PA (labeled "MAIN") and the other containing the envelope detection and processing circuits (labeled "ENV"). Table 9.2 lists the material used in this setup.

The 2-tone signal is generated using the RF signal generator SMIQ06B and the function generator HP33250A. The function generator is set to output a sinewave

Table 9.2 List of material used in the 2-tone measurement setup

Material	Description
SMIQ06B [16]	RF signal generator
FSIQ7 [15]	RF signal analyzer
LC534AM [12]	Oscilloscope
HP33250A [4]	Function generator
E018-06.D [8]	DC voltage source
Fluke 45 [9]	Bench-top digital multimeter
Fluke 85 [10]	True-rms handheld digital multimeter
CP015 [13]	Current probe
K251 [6]	Coaxial bias-tee
ZN2PD-9G [14]	Coaxial power splitter
SUCOFLEX 104 [11]	Coaxial cable (SMA)

with a frequency equal to half of the desired tone spacing, 0 V offset, and an amplitude that is varied so that the power of the tones can be swept. The output of the function generator is connected to the "I" input of the RF signal generator. The RF signal generator is configured to output a vector modulated signal at the desired RF frequency. In this vector modulation, the signal at its "I" input modulates the RF carrier—which is set for the maximum power and kept fixed—producing at the output an amplitude modulated signal whose spectrum consists of two tones centered at the RF carrier frequency and having a space between them equal to double the modulation frequency (function generator signal frequency). This 2-tone spectrum is the same of that obtained from a double sideband suppressed carrier (DSBSC) AM system [7, Chap. 9]. This strategy for generating 2-tone signals is useful when two RF signal generators are not available.

The 2-tone signal is split in two signals using a coaxial power splitter. These signals are fed to the PA and to the envelope detector. The processed envelope from the ENV board is provided to the MAIN board through a coaxial SMB cable. The total current consumed by the PA and modulator comes from the VDD power supply and are measured with the true-rms handheld digital multimeter. The dynamic supply signal is monitored with the LC534AM oscilloscope which is also used to monitor the current being consumed by the PA through the clamp-on current probe CP015. The RF output is connected to the signal analyzer FSIQ7 and both the output power and the IMD3 are measured.

Essentially the same setup is used in the 2-tone measurements for the frequency-tunable system characterization. The main differences are that just one evaluation board is used, no power splitter is needed, and the PA supply voltage and current do not need to be monitored with an oscilloscope.

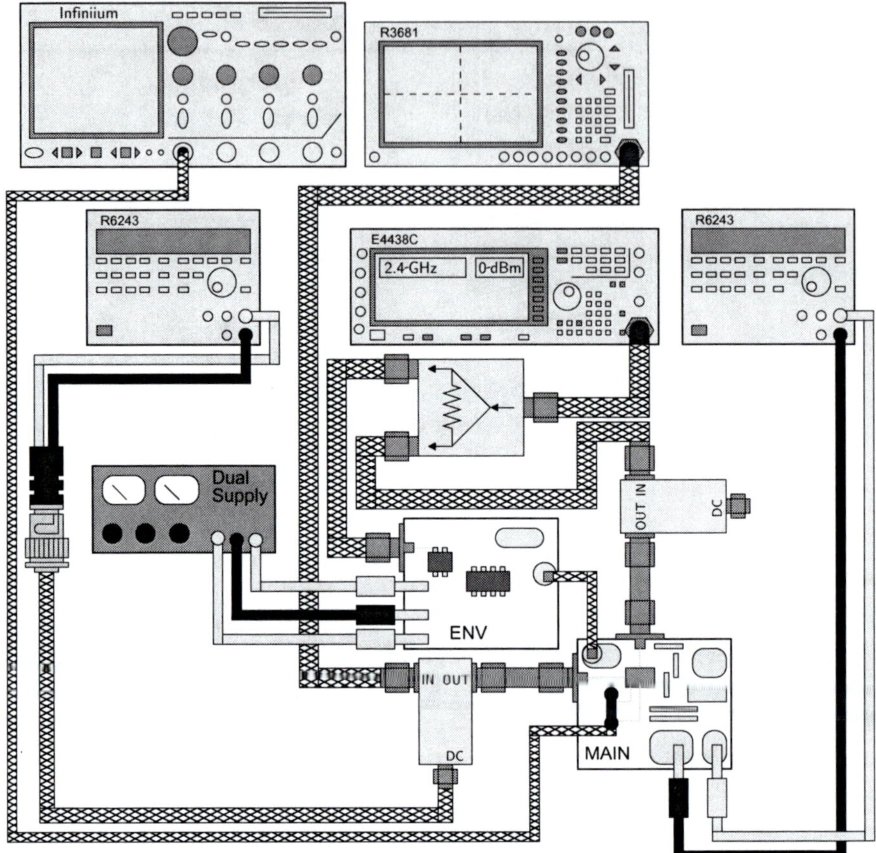

Fig. 9.4 Setup for OFDM measurements

9.3 Setup Used for the OFDM Measurements

Figure 9.4 depicts the measurement setup used for the OFDM characterization. This setup was used for the measurements presented in Sect. 4.5.2. These measurements were performed in the Fujitsu Laboratories facilities in Kawasaki, Japan.

The OFDM signal is provided by the signal generator E4438C which is split and fed to the MAIN and ENV evaluation boards. The amplified signal is characterized with the signal analyzer R3681 which provides the output power as well as the EVM readings. The oscilloscope Infiniium is used to monitor the dynamic supply signal. The total current consumed by the RF power amplifier and modulator is read from the DC source that also functions as a current monitor. The list of material used in this setup is given in Table 9.3.

Table 9.3 List of material used in the OFDM measurement setup

Material	Description
E4438C [5]	Signal generator
R3681 [2]	Signal analyzer
R6243 [1]	DC power supply and monitor
K251 [6]	Coaxial bias-tee
ZN2PD-9G [14]	Coaxial power splitter
SUCOFLEX 104 [11]	Coaxial cable (SMA)

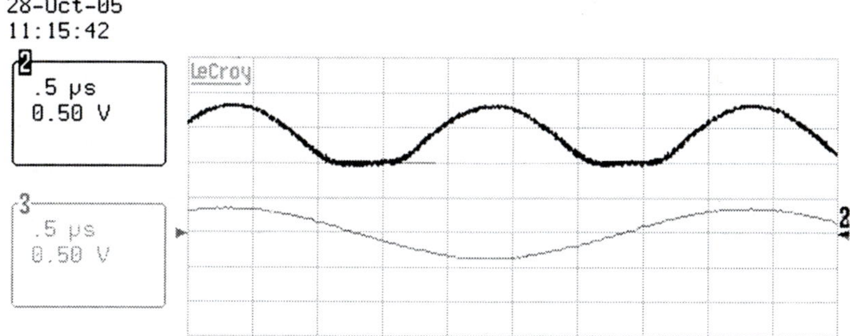

Fig. 9.5 Dynamic supply signal for 2-tone excitation

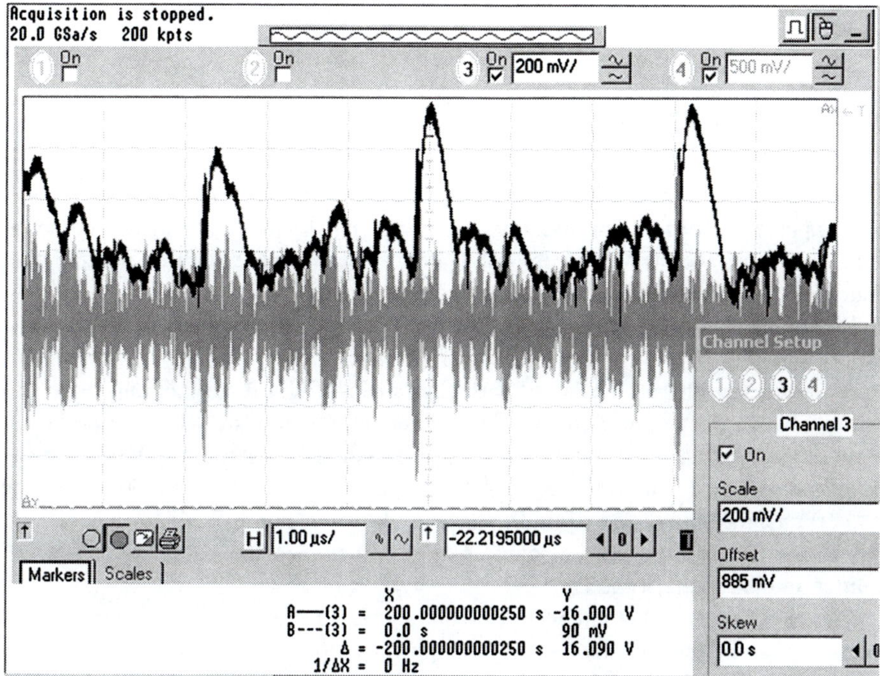

Fig. 9.6 Dynamic supply signal for OFDM excitation showing envelope detection delay

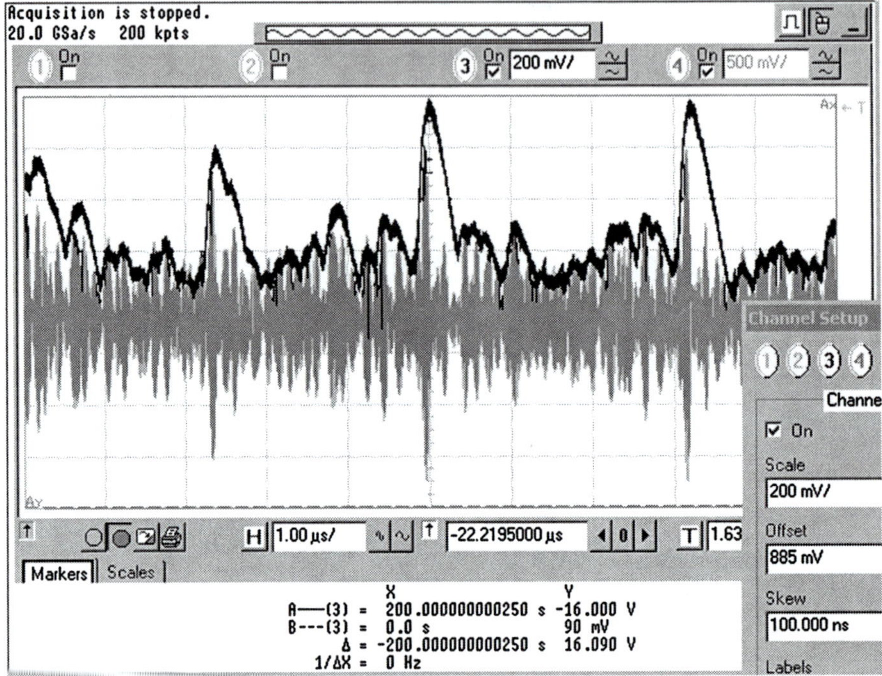

Fig. 9.7 Dynamic supply signal for OFDM excitation. Delay of the envelope detection compensated by the oscilloscope

9.4 Additional Screenshots

Figure 9.5 shows a screenshot of the oscilloscope LC534AM with two waveforms. The upper waveform is the dynamic supply signal for a 2-tone measurement at 5.2 GHz. The waveform at the bottom is the 250 kHz signal from the function generator HP33250A used to generate the two input tones.

Figure 9.6 shows the screenshot of the oscilloscope Infiniium where the dynamic supply signal (*top*) and the input RF signal (*bottom*) can be observed. These signals were obtained with an OFDM measurement at 2.4 GHz. It reveals that there is a delay between the two waveforms. This delay already appears when we probe the input of the modulator with the oscilloscope. It indicates that the origin of the delay is in the envelope detector. Using the deskew function of the oscilloscope and setting it to 100 ns, the signals are again in phase as shown in Fig. 9.7. To compensate for this delay, a very long coaxial cable was used between the output of the power splitter and the input of the PA.

Figure 9.8 shows the screenshot of the signal analyzer R3681 for the OFDM measurement of the dynamic supply RF power amplifier at 2.4 GHz. The output power is 11 dBm (12.6 mW) and the rms value of the EVM is 3%. The constellation of the output signal can be seen at the bottom right corner of the figure.

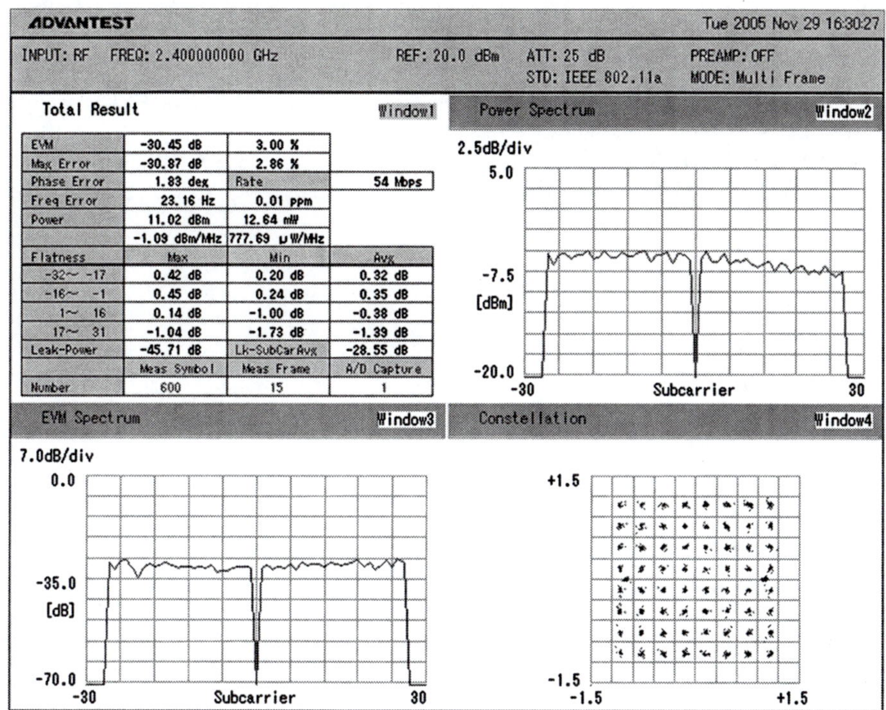

Fig. 9.8 OFDM output signal characteristics from signal analyzer

References

1. Advantest (2002) R6243/R6244 DC voltage current source/monitor. Data Sheet. URL http://www2.rohde-schwarz.com/file/R6243_1E.pdf
2. Advantest (2010) R3671/R3681 signal analyzers. Catalog. URL http://www.advantest.co.jp/en-index.shtml
3. Agilent (1999) 8719D network analyzers. Agilent Technologies, USA. URL http://cp.literature.agilent.com/litweb/pdf/08720-90288.pdf
4. Agilent (2003) 33250A 80 MHz Function/Arbitrary waveform generator. Agilent Technologies, USA. URL http://cp.literature.agilent.com/litweb/pdf/33250-90002.pdf
5. Agilent (2006) E4438C ESG vector signal generator. Data Sheet. URL http://cp.literature.agilent.com/litweb/pdf/5988-4039EN.pdf
6. Anritsu (2000) Ultra-wideband bias tees—K251 and V251. Data Sheet. URL http://www.eu.anritsu.com/files/11410-00253.pdf
7. Cripps SC (2006) RF Power Amplifiers for Wireless Communications, 2nd edn. Artech House, Norwood
8. Delta (2001) E 018-0.6D ±0–18 V, 0.6 A regulated dual power supply. Delta Elektronika, Zierikzee, The Neatherlands. URL http://www.delta-elektronika.nl/Museum.htm
9. Fluke (1989) 45—Dual display multimeter. Fluke, Everett, WA. URL http://www.fluke.com
10. Fluke (2003) 80 Series III multimeter. Fluke, Everett, WA. URL http://www.fluke.com
11. Huber+Suhner (2005) Coaxial cable: SUCOFLEX_104—Device IDs: 240982/4, 240983/4, 240985/4, 240986/4. Data Sheet. URL http://www.hubersuhner.com
12. LeCroy (1999) LC series color digital oscilloscopes. LeCroy, Chestnut Ridge, NY. URL www.lecroy.com

13. LeCroy (2001) CP015. LeCroy, Chestnut Ridge, NY. URL http://www.lecroy.com
14. Mini-Circuits (2008) ZN2PD-9G—2 way-0°, 50 Ω, 1700 to 9000 MHz power splitter/combiner. Data Sheet. URL http://www.mini-circuits.com/pdfs/ZN2PD-9G.pdf
15. Rohde & Schwarz (2002) FSIQ7 signal analyzer. Rohde & Schwarz, Germany. URL http://www.rohde-schwarz.com
16. Rohde & Schwarz (2002) SMIQ06B vector signal generator. Rohde & Schwarz, Germany. URL http://www.rohde-schwarz.com

Chapter 10
Appendix B: Procedure for Impedance Matching of Printed-Circuit RF Amplifiers

Abstract This second and last appendix presents a systematic procedure for impedance matching of RF amplifiers mounted on printed-circuit boards. This procedure involves two-tier de-embedding techniques and on-board impedance matching aided by the Smith chart. The application of this procedure to match a CMOS RF power amplifier operating at 5.2 GHz resulted in a power gain of 8 dB, which is 5.6 dB higher than its unmatched gain (2.4 dB) and only 1.1 dB lower than its maximum theoretical gain (9.1 dB).

10.1 Introduction

Impedance matching is an important issue in the design and characterization of an RF amplifier. When the RF amplifier is mounted on a printed-circuit board (PCB), impedance matching can constitute a difficult and painstaking task. If the amplifier is a custom integrated circuit, the difficulties are associated mainly with the incorrect estimation of the integrated circuit parasitics, like pad capacitance and bondwire inductance. If the amplifier is based on a commercial transistor, the problem may be the lack of information about the transistor parameters at the operating frequency.

The process of matching a printed-circuit RF amplifier can be divided in two main parts. The first consists of the application of two-tier de-embedding techniques to determine the scattering parameters (S parameters) of the device under test (DUT). In this case, the DUT is the integrated amplifier or the discrete transistor that implements the amplifier. The second part is the implementation of the input and output impedance matching networks on the PCB.

Well known two-tier de-embedding techniques were developed about three decades ago [8, 14, 16, 27] and have been intensively employed and improved since then [5, 12, 13, 19, 21, 30–35]. The subject of impedance matching has complete chapters [10, 17, 20, 26] or entire books [1] dedicated to it. Although the literature

P.A. Dal Fabbro, M. Kayal, *Linear CMOS RF Power Amplifiers for Wireless Applications,*   137
Analog Circuits and Signal Processing,
DOI 10.1007/978-90-481-9361-5_10, © Springer Science+Business Media B.V. 2010

is exhaustive for both parts of the matching process, only a few papers deal with the two of them together [9, 23, 24].

O'Reilly, Neidert, and Wilson in [24] proposed a computer-aided procedure for RF power amplifier matching in which an optimization routine provides the dimensions of the microstrip sections for the on-board matching. The drawback of their procedure is that the routine requires the knowledge of the impedance to be presented at the input and output of the amplifiers. These impedances are determined with conductive tape overlays that are adjusted so that the desired performance can be achieved.

Nickel and Schutt-Ainé in [23] presented a methodology for narrowband amplifier design where chip capacitors are used to form matching networks in a coupled microstrip system. The position and value of the chip capacitors are optimized via software. The disadvantage of their approach is that the parameters of the transistor are obtained through measurements using a separate fixture.

Choi, Youm, and Hwang in [9] developed an S-parameter extraction method for PCB mounted RF amplifiers. The S parameters of the extracted transistor (or amplifier) are then used to design the impedance matching network. The drawback of their procedure is that the design of an additional PCB to extract the transistor S parameters is required.

A common drawback of these three approaches is the necessity of a dedicated fixture to determine the value of a fundamental variable. In [24], this variable is the impedance to be presented to the input/output of the amplifier, whereas in [9, 23] it is the set of S parameters of the amplifier.

With the objective of eliminating the need of a second board (or fixture), we developed a new procedure for systematic impedance matching of printed-circuit RF amplifiers. In this new procedure, no optimization is required, and the information obtained with two-tier de-embedding techniques is used to implement on-board matching networks assisted by the Smith chart.

The procedure is divided in five steps and a detailed description for each of them is presented. Summarized guidelines are given for quick reference. A practical example illustrates the application of these guidelines in the impedance matching of a chip-on-board CMOS power amplifier operating at 5.2 GHz. The results are discussed and compared with previous works.

10.2 Procedure for Impedance Matching

Consider the characterization of an integrated RF power amplifier[1] bonded to a printed-circuit evaluation board using the chip-on-board technique. The access to the input and output terminals of this amplifier is made through microstrips that ar-

---

[1]It could also be a commercial RF amplifier or transistor, but, in these cases, the matching difficulties could be caused by the lack of information about package parasitics at the desired frequency.

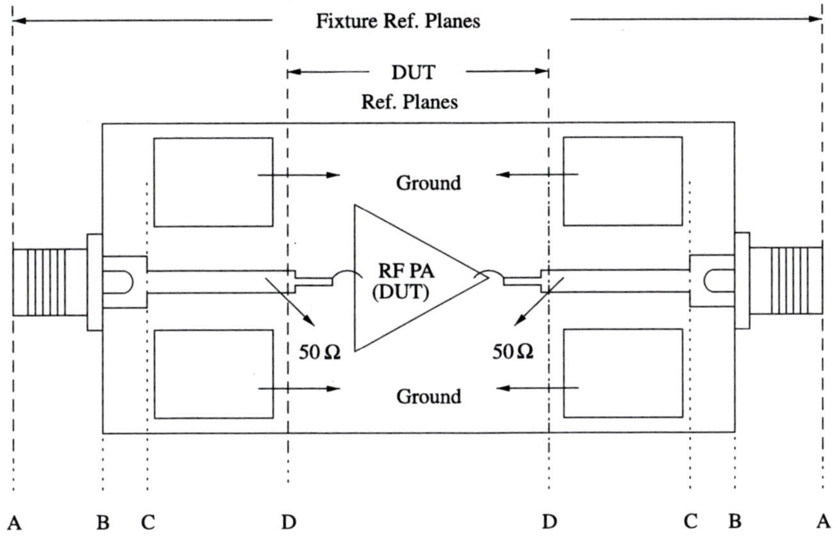

Fig. 10.1 PCB test fixture for an integrated RF power amplifier (PA)

rive to the edges of the board and provide microstrip-to-coaxial interfaces through
SMA connectors. A possible PCB fixture for this purpose is depicted in Fig. 10.1.
If the integrated circuit parasitics including pads and bondwires were not correctly
estimated in the design phase, or if the impedance matching networks were chosen
to be partially or completely placed off-chip, the process of characterization can be
cumbersome. Using two-tier de-embedding techniques, the impedance at the DUT
reference planes can be determined. However, the knowledge of these impedances
does not solve the problem of input and output matching. At this point, overlaying
pieces of conductive tape to form matching sections [24] can, by trial-and-error,
lead to the desired matching. Another possibility is to fabricate a new PCB with
new impedance matching networks designed based on the extracted parameters of
the DUT [9, 23]. However, the procedure described in this appendix shows that the
fabrication of an extra PCB can be avoided.

The procedure consists of five major steps:

1. PCB access lines design.
2. Short-Open-Load-Thru (SOLT) calibration and measurement.
3. Thru-Reflect-Line (TRL) calibration and measurement.
4. Access line modeling and de-embedding.
5. Matching network implementation.

The flowchart describing this procedure is depicted in Fig. 10.2. Sections 10.2.1
to 10.2.5 provide detailed explanation for each of the five steps. The same organiza-
tion is used in Sect. 10.3.

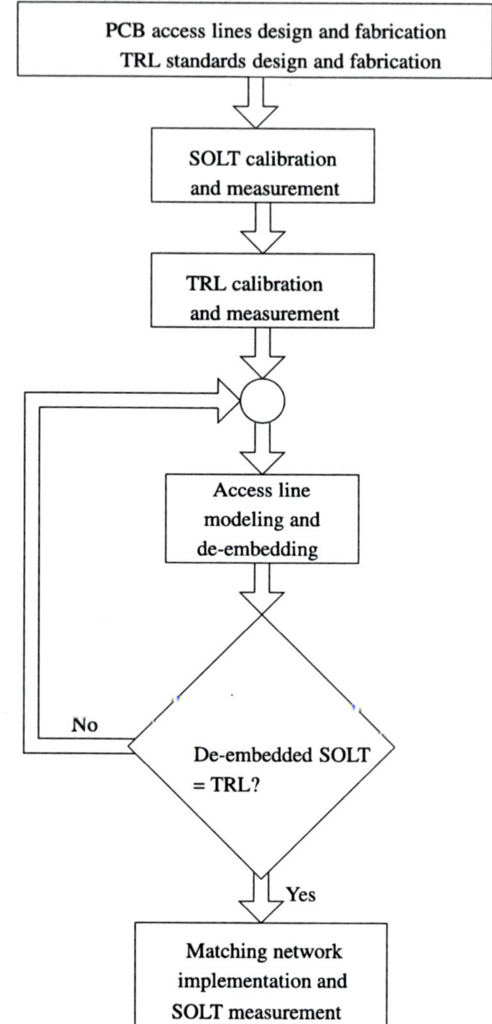

Fig. 10.2 Impedance matching procedure flowchart

10.2.1 PCB Access Lines Design and Fabrication

The microstrip lines at the input and output of the amplifier are very important and must be carefully designed. They should have a characteristic impedance of 50 Ω for measurements with standard network analyzers. Furthermore, the length should be at least $\lambda/2$ so that any kind of impedance can be matched with a shunt capacitor [18]. However, it should be as short as possible to reduce losses. It is important to note that the TRL standards (refer to Sect. 10.2.3) should also be designed and fabricated simultaneously with the evaluation PCB.

10.2.2 SOLT Calibration and Measurement

To achieve the desired input and output match in the characterization of an RF power amplifier, the measurement of its S parameters is mandatory. A network analyzer is required for this purpose. Calibration of the network analyzer is a part of well known one- and two-tier de-embedding techniques [13, 19, 32] and is necessary to correct the systematic errors of the measurement system [15]. In the procedure described in this appendix, calibration and de-embedding are treated as separate actions. The term *de-embedding* is restricted to the sense used by Bauer and Penfield in [8]:

> De-embedding is the process of deducing the impedance of a DUT from measurements made at a distance, when the electrical properties of the intervening structure are known.

The SOLT technique [4] is a full two-port calibration procedure that uses the short, open, load, and thru standards to remove all the twelve correctable systematic errors from an S-parameters measurement [3]. Although these standards are required to be fully characterized [12], they are usually provided by the network analyzer manufacturer. The SOLT calibration shifts the measurement reference plane from the network analyzer connectors to the SMA connectors in the fixture, thereby removing the effect of the elements between the network analyzer and the PCB fixture such as cables, bias-tees, and DC blocks. In this first step, an SOLT calibration is performed, the circuit is measured and the resulting data are stored.

10.2.3 TRL Calibration and Measurement

The SOLT calibration provides measurement data in which the effect of the PCB fixture—SMA connectors and microstrip lines—is still present. These elements, following the nomenclature used in [8, 32, 35], will be referred hereafter as to embedding networks. The TRL method [4] used in this step shifts the measurement reference planes to the DUT planes (see Fig. 10.1), thereby providing information about the actual input and output impedances of the DUT—free of the fixture effects. The standards used in the TRL calibration—thru, reflect, and line—must be fabricated according to the fixture being used. However, these standards are not required to be fully characterized [35], which make the TRL calibration to suit well to PCB fixture de-embedding. The TRL calibration is based on the 8-term error model [6, 29] to remove the measurement errors. It was proposed by Engen and Hoer in [14] and represents an improvement in relation to the thru-short-delay technique introduced by Franzen and Speciale in [16] because the need for a short of known reflection coefficient was eliminated.

The TRL standards and the evaluation PCB should be fabricated simultaneously. For the thru standard, the input and output embedding networks are connected directly to each other providing a zero-length thru. For the reflection, an open or a short can be used (by leaving the microstrip lines of the embedding network open or connected to ground), whereas a $\lambda/4$ transmission line with a characteristic impedance of 50 Ω connecting the input and output embedding networks can play the role of the line standard. After the calibration has been performed on the network analyzer using these custom standards, the S parameters of the circuit are measured and the resulting data are stored.

10.2.4 Access Line Modeling and De-Embedding

After the second and third steps, information about the impedances at the fixture reference planes and the DUT reference planes is available. This fourth step will provide information about the embedding networks, which includes the SMA connector and the microstrip lines linking the connector to the amplifier input/output, as depicted in Fig. 10.1. This process of determining the characteristics of the intervening fixture is known as unterminating [8, 35].

The model for the microstrip lines can be built by directly using the PCB layout dimensions and the PCB laminate properties, provided that the laminate has tight tolerances—which is the case of commercially available RF laminates. A simple electrical delay can be used to model the effect of the SMA connector. This delay can be found by a direct reflection measurement of the SMA connector with its tab shorted to its panel. Once the embedding network has been modeled it can be de-embedded from the SOLT scattering parameters measurement. The resulting S parameters can then be compared with the TRL data, which must closely agree. If unacceptable discrepancies exist, the embedding network model parameters can be optimized until a good agreement is reached.

10.2.5 Matching Network Implementation

With the knowledge of the embedding network parameters and of the impedances at the fixture and DUT reference planes granted by the previous steps, the task of designing an impedance matching network is almost accomplished. In this step, to ensure that the desired matching is achieved, the embedding network model is added in cascade to the TRL measurement data (DUT reference plane)—both at the input and output—to recover the S parameters at the fixture reference plane.

With the aid of a Smith chart, shunt capacitors to ground can be virtually added to the access line to form π- or T-networks. Hence, the position and the value of the capacitors that result in the desired input and output impedance matching can be known prior to their physical placement on the PCB. At frequencies close to the self-resonant frequency of the chip capacitor, one must be aware that the apparent capacitance may be above its nominal value and, therefore, its careful choice is very important. The capacitors are then soldered on the evaluation board.

10.2.6 Summarized Guidelines

The list below summarizes the procedure described in this section and can be used together with the flowchart of Fig. 10.2 as a quick reference guide.

1. Design and fabricate the PCB fixture with special attention to the input and output microstrip access lines. TRL standards should be fabricated simultaneously.
2. Make SOLT calibration and measurement.
3. Make TRL calibration and measurement.
4. a. Model the fixture elements (embedding network).
   b. De-embed the fixture elements from the SOLT measurement data.
   c. Compare the resulting data with TRL measurement data to validate the embedding network model.
5. a. Add the embedding network model to the TRL measurement data.
   b. Virtually add capacitor(s) in shunt to the microstrip access lines in a position that results in the desired matching.
   c. Physically place the capacitor(s) on the PCB.
   d. Make final SOLT measurement.

10.3 Application Example

To illustrate the procedure for impedance matching just described, this section presents a real and practical example that involves the characterization of a 5.2 GHz single-stage RF power amplifier integrated in a 0.11 μm CMOS technology.

10.3.1 First Step

The printed-circuit board for the evaluation of the RF PA is designed taking care that a 50 Ω transmission line of a length of at least $\lambda/2$ is present in the input and output access lines from the SMA connectors to the input and output of the PA. The chip-on-board method is used to mount the integrated PA onto the board. Hence, the SMA connectors, the launchers onto which the connectors are soldered, and the 50 Ω microstrip line form the access lines. In both sides of the 50 Ω microstrip lines, ground planes are created to facilitate the placement of 0402-case-size chip capacitors in shunt to them. The possible PCB fixture shown in Fig. 10.1 is a representation (not to scale) of the real fixture used in this example. The TRL standards and the PCB were designed and fabricated at the same time.

10.3.2 Second Step

Figures 10.3 and 10.4 show the small signal S parameters S_{11} (input reflection coefficient) and S_{22} (output reflection coefficient), and Fig. 10.5 shows S_{21} (forward

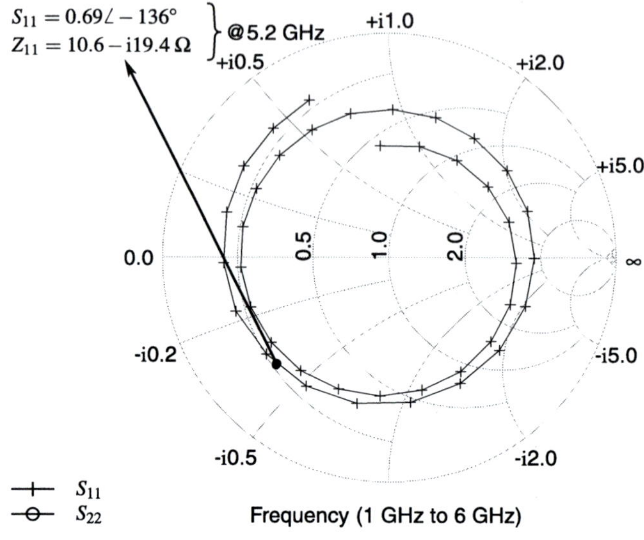

Fig. 10.3 Unmatched power amplifier SOLT measurement (S_{11})

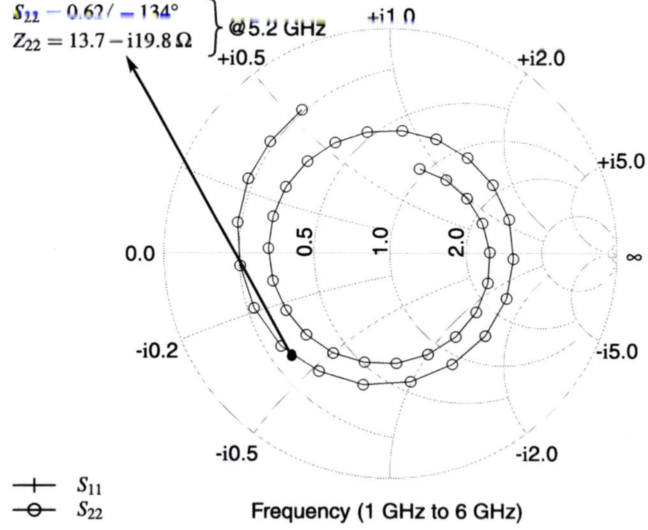

Fig. 10.4 Unmatched power amplifier SOLT measurement (S_{22})

power gain) and S_{12} (reverse power gain) measured at the fixture reference planes (refer to Fig. 10.1) with the HP8719D network analyzer [3] after being calibrated with the SOLT standards [7] provided by the manufacturer. These S parameters are hereafter called S_{solt}.

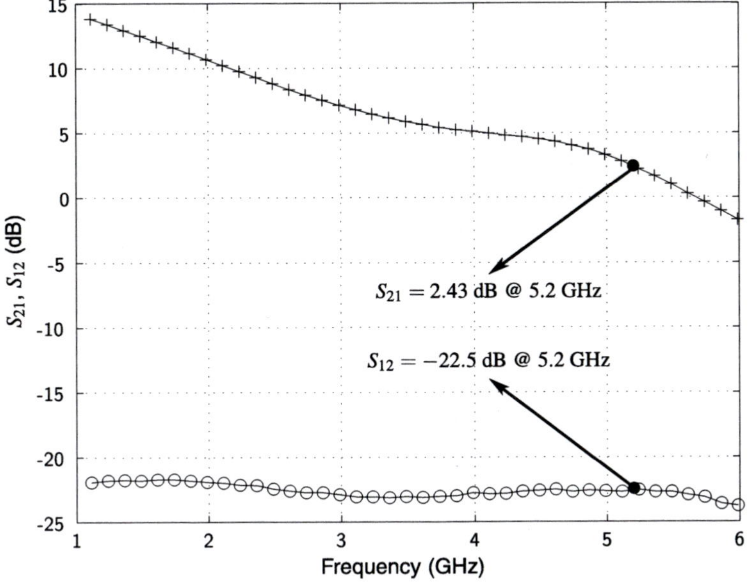

Fig. 10.5 Unmatched power amplifier SOLT measurement (S_{21} and S_{12})

Table 10.1 Summary of the measured SOLT S and Z parameters for the unmatched power amplifier at 5.2 GHz

Parameter	Value	Parameter	Value
S_{11}	$0.69\angle - 136°$	Z_{11} (Ω)	$10.6 - i19.4$
S_{22}	$0.62\angle - 134°$	Z_{22} (Ω)	$13.7 - i19.8$
S_{11} (dB)	-3.22	S_{21} (dB)	2.43
S_{22} (dB)	-4.15	S_{12} (dB)	-22.5

For this example, it is considered that a conjugate match at both input and output is desired so that the amplifier can attain the maximum power gain. From Figs. 10.3 and 10.4, it can be noticed that the power amplifier is unmatched at 5.2 GHz, resulting in a power gain of 2.4 dB at this frequency (see Table 10.1 for a summary of these results). The objective, hence, is to bring the markers at this frequency the closest possible to the center of the Smith chart, where a perfect match takes place.

10.3.3 Third Step

The network analyzer is then calibrated with the TRL standards fabricated in step 1. The measured S parameters (S_{trl}) are now referenced to the DUT reference planes (refer to Fig. 10.1) and the results are shown in Fig. 10.6. Comparison of this figure with Figs. 10.3 and 10.4 reveals that, at 5.2 GHz, the embedding networks change

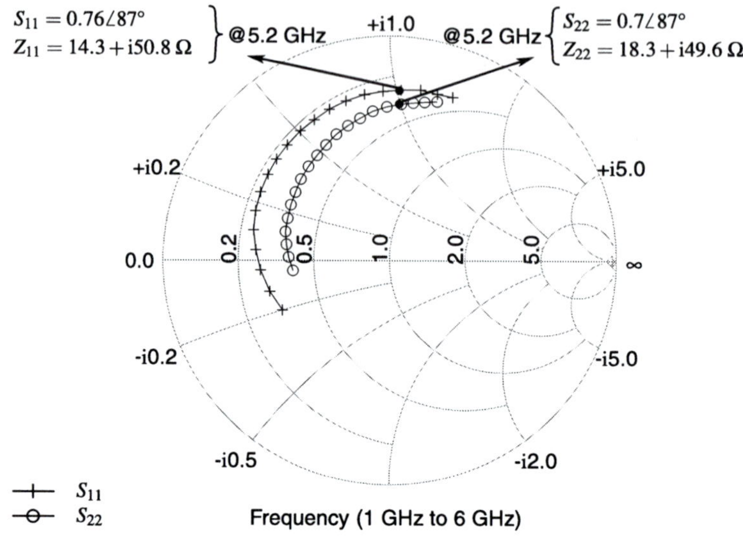

Fig. 10.6 Unmatched PA TRL measurement (S_{11} and S_{22})

significantly the impedance at the input and output of the power amplifier (integrated circuit pads, bondwires, and the bondwire receptacles in the PCB are part of the DUT).

10.3.4 Fourth Step

The model for the microstrip access lines for the input and output of the power amplifier is built based on the PCB layout data without the need for fine tuning the model. This is possible because the substrate used in the PCB is an RF laminate with tight tolerances (RO4350 [28]). It is easier if the input and output access lines are exactly equal so that just one model is necessary (and just one set of TRL standards is required).

Figure 10.7 depicts the model as an ensemble of transmission lines. The sections A–B, B–C, and C–D indicated in this figure correspond to the sections of equal names in Fig. 10.1. The parameters used in the model are the characteristic impedance (Z_0), width (W), and length (L) of the transmission lines; and the relative permittivity (ε_r), thickness (T), and dielectric loss ($\tan\delta$) of the laminate. The electrical delay of the flange-mount SMA connector used in this example was measured to be $t_{ed} = 40$ ps. This was converted into the electrical length $l = 12$ mm of the ideal 50 Ω transmission line used in the model (A–B). This conversion uses the phase velocity of light in a medium of effective permittivity $\varepsilon_e = 1$ [26], according to the following relationship:

$$t_{ed} = \frac{l \times \sqrt{\varepsilon_e}}{c}. \tag{10.1}$$

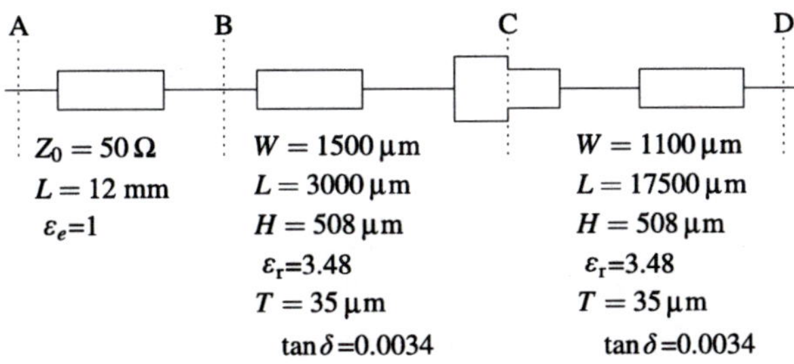

Fig. 10.7 Embedding network model

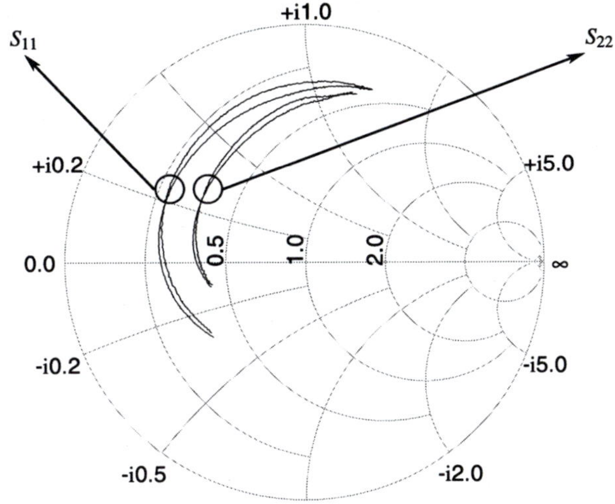

Fig. 10.8 Unmatched PA—S_{11}, S_{22} comparison between TRL and de-embedded SOLT

The model of the input embedding network can be seen as a two-port network with port 1 and port 2 located, respectively, at the A and D planes in Fig. 10.7 (and in Fig. 10.1). For the output embedding network, port 1 and port 2 are located, respectively, at the D and A planes. Hence, the S parameters S_{in} and S_{out} of these embedding networks can be determined. S_{in}, S_{out}, and S_{solt} are converted into the chain scattering parameters [17] T_{in}, T_{out}, and T_{solt} to extract the S parameters of the DUT from the SOLT measurement data. The extracted chain S parameters of the DUT are obtained by using (10.2), which are then converted into S parameters (S_{solt_deemb}).

$$[T_{solt\_deemb}] = [T_{in}]^{-1} \times [T_{solt}] \times [T_{out}]^{-1}. \tag{10.2}$$

In this example, Agilent's Advanced Design System (ADS) [2] was used to facilitate the de-embedding procedure. It incorporates an easy-to-use de-embedding block based on the principle described in [8, Fig. 2].

After de-embedding, S_{solt_deemb} can be compared with S_{trl}, which should closely agree since both correspond to the S parameters referenced to the DUT planes (refer to Fig. 10.1). S_{11} and S_{22} for these two sets of S parameters are shown in Fig. 10.8,

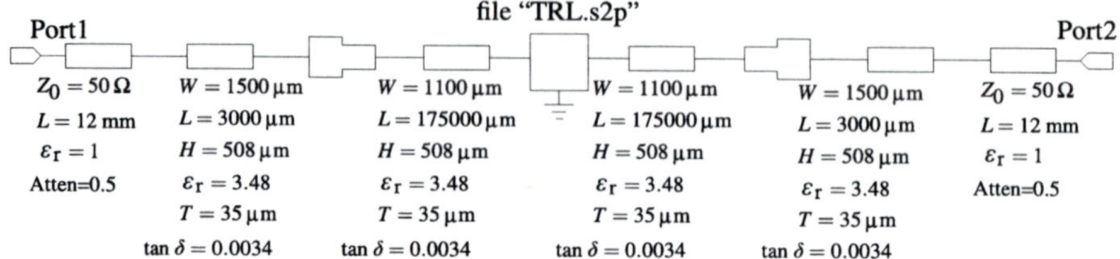

Fig. 10.9 Schematic showing the TRL data cascaded with the input and output embedding networks

revealing that a good agreement exists between them. If the agreement is not satisfactory, the components that compose the embedding network model can be fine tuned to minimize the discrepancies.

Once the agreement is found to be enough, the embedding network data is added in cascade to the TRL data (or de-embedded SOLT data) to reconstitute the original SOLT measurement data, but at this time with complete knowledge of the embedding network parameters and control of where to place shunt capacitors. The situation at schematic level is depicted in Fig. 10.9.

10.3.5 Fifth Step

As previously stated, this example targets a simultaneous conjugate match at the input and output of the power amplifier at an operating frequency of 5.2 GHz. For this kind of matching, it is required that $\Gamma_{\mathrm{S}} = \Gamma_{\mathrm{IN}}^{*}$ and $\Gamma_{\mathrm{L}} = \Gamma_{\mathrm{OUT}}^{*}$, where Γ_{S}, Γ_{IN}, Γ_{L}, and Γ_{OUT} are, respectively, the source, input, load, and output reflection coefficients. The source (Γ_{MS}) and load (Γ_{ML}) reflection coefficients for a simultaneous conjugate match can be calculated based on the unmatched S parameters of the amplifier according to the equations presented in [17] and repeated here for clarity:

$$\Gamma_{\mathrm{MS}} = \frac{B_1 \pm \sqrt{B_1^2 - 4|(S_{11} - \Delta S_{22}^{*})|^2}}{2(S_{11} - \Delta S_{22}^{*})}, \tag{10.3}$$

$$\Gamma_{\mathrm{ML}} = \frac{B_2 \pm \sqrt{B_2^2 - 4|(S_{22} - \Delta S_{11}^{*})|^2}}{2(S_{22} - \Delta S_{11}^{*})}, \tag{10.4}$$

$$B_1 = 1 + |S_{11}|^2 - |S_{22}|^2 - |\Delta|^2, \tag{10.5}$$

$$B_2 = 1 + |S_{22}|^2 - |S_{11}|^2 - |\Delta|^2, \tag{10.6}$$

where Δ is the determinant of the two-port amplifier S-parameters matrix: $\Delta = S_{11}S_{22} - S_{12}S_{21}$.

The source (Z_{MS}) and load (Z_{ML}) impedances to be presented at the amplifier input and output are determined using (10.3) and (10.4). The maximum transducer

Table 10.2 Required values of source and load impedances for simultaneous conjugate match and corresponding maximum gain at 5.2 GHz

Parameter	Value
Z_{MS} (Ω)	$8 - i47.4$
Z_{ML} (Ω)	$9.4 - i43.3$
G_{max} (dB)	9.1

power gain under simultaneous conjugate match is calculated using S_{trl} and the expression below [17].

$$G_{max} = \frac{1}{1 - |\Gamma_{MS}|^2}\left(|S_{21}^2|\frac{1 - |\Gamma_{ML}|^2}{|1 - S_{22}\Gamma_{ML}|^2}\right). \tag{10.7}$$

Table 10.2 shows the values found after applying (10.3)–(10.7) to the S parameters of the TRL measurement at 5.2 GHz.

The maximum theoretical gain is 9.1 dB. Comparing to the measured unmatched power gain of 2.4 dB given in Table 10.1, there is a large improvement margin to be exploited through impedance matching.

The embedding network model developed previously allows one to determine the value and position of shunt capacitors that can be added to the microstrip access lines to attain the source and load impedance values of Table 10.2. This can be done before physically placing the capacitors on the board, avoiding trial-and-error impedance matching implementation. The Smith chart can be very useful in the determination of the position and value of these capacitors. The software Smith Chart presented in [25], whose demo version can be downloaded from [11], is used in this example for this purpose. By choosing the frequency of interest and beginning from the 50 Ω impedance of the network analyzer ports, the change to this impedance caused by the embedding networks can be traced on a Smith chart for every micron of the transmission lines. This gives the designer the ability to exactly know where to place the chip capacitors and which values must be chosen.

10.3.5.1 Matching the Output

The departing point is 50 Ω and the objective is to transform it in the desired load impedance given in Table 10.2: $Z_{ML} = 9.4 - i43.3$ at 5.2 GHz. Using the Smith Chart tool and referring to Fig. 10.10, the departing point is marked as number **1**. Then, the pieces of transmission line that compose the output embedding network are added according to the model of Fig. 10.7.[2] Point **2** is superposed to Point **1** because the SMA connector between these two points was modeled as a 50 Ω line with an electrical delay. Point **4** is superposed to Point **3** because the 50 Ω microstrip line between these two points has a length of $\lambda/2$ and, hence, it provides a complete turn around the center of the Smith chart. The impedance of $44.5 - i7.5$ Ω at point **4** is the load impedance seen by the unmatched power amplifier.

---

[2]In this example, without loss of accuracy, the step discontinuities in the microstrip line model were considered to be lossless when using the Smith Chart software.

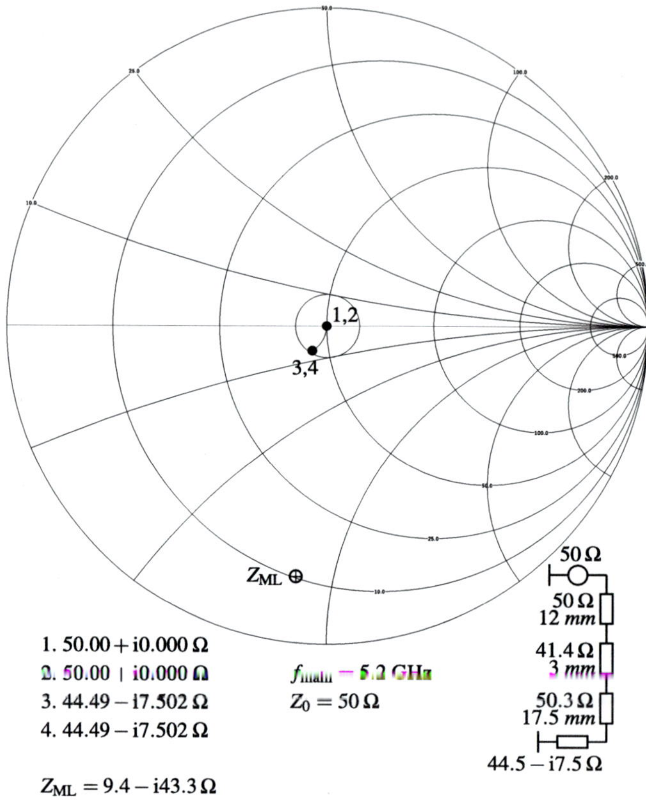

Fig. 10.10 5.2 GHz load impedance transformation caused by the embedding network. Target load impedance also shown

The target load impedance is also depicted in the figure. With a 1.8 pF capacitor in shunt with the 50 Ω microstrip between points **3** and **4**, the load impedance gets close to the target value as shown in Fig. 10.11.

Operation at 5.2 GHz requires special attention to the apparent capacitance that the discrete capacitors present at this frequency. The S parameters provided by the manufacturer [22] show that, at 5.2 GHz, a 1 pF chip capacitor has the apparent capacitance the closest to 1.8 pF among the available components (other options were 1.5 pF, 1.8 pF, and 2.2 pF). This chip capacitor was chosen for the output matching network.

10.3.5.2 Matching the Input

At the input, the departing point is also 50 Ω and the objective is to transform it in the desired source impedance given in Table 10.2: $Z_{MS} = 8 - i47.4$ at 5.2 GHz. Using again the Smith Chart tool, the departing point is depicted as point **1** in Fig. 10.12

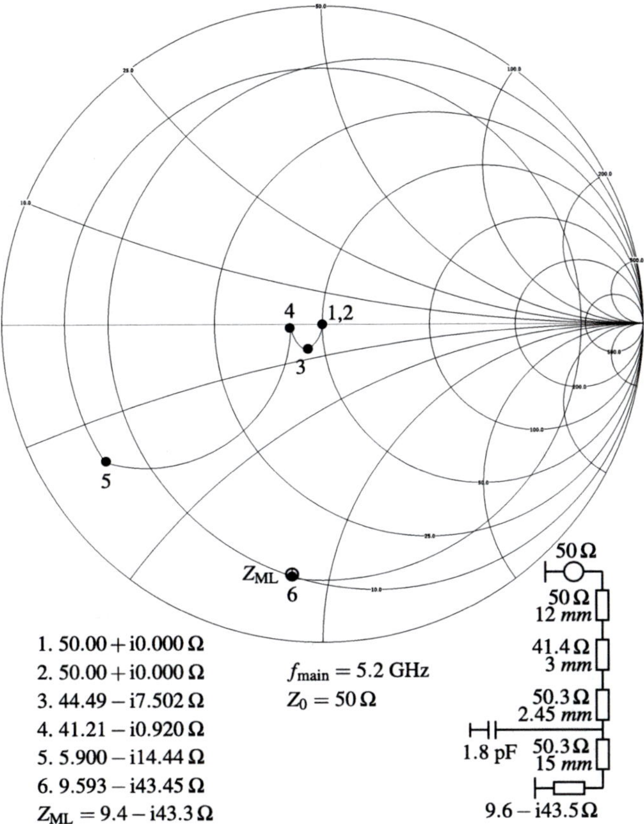

1. $50.00 + i0.000 \, \Omega$

2. $50.00 + i0.000 \, \Omega$

3. $44.49 - i7.502 \, \Omega$ $f_{\text{main}} = 5.2 \text{ GHz}$

4. $41.21 - i0.920 \, \Omega$ $Z_0 = 50 \, \Omega$

5. $5.900 - i14.44 \, \Omega$

6. $9.593 - i43.45 \, \Omega$

$Z_{ML} = 9.4 - i43.3 \, \Omega$

Fig. 10.11 Shunt capacitor added to the output microstrip access lines between points **3** and **4** of the previous figure

and, as the input embedding network is exactly the same as that of the output, the transformation is the same. Therefore, Fig. 10.12 is equal to Fig. 10.10 with the only difference that in Fig. 10.12 the target source impedance Z_{MS} is shown instead of Z_{ML}.

According to Fig. 10.12, a shunt capacitor of 2.1 pF at a position between points **3** and **4** brings the source impedance to the desired value. The effect of the addition of this capacitor is observed in Fig. 10.13.

At 5.2 GHz, a 1 pF chip capacitor [22] presents an apparent capacitance closer to 2.1 pF than the other values that were available. Hence, this was also the value chosen for the chip capacitor for the input matching network.

After the values of the capacitors and their positions have been determined, ADS was used to include the model of the capacitors provided by the manufacturer and verify that the expected matching was achieved.[3] The components were then sol-

---

[3]This can also be done through matrix algebra making optional the use of the simulator.

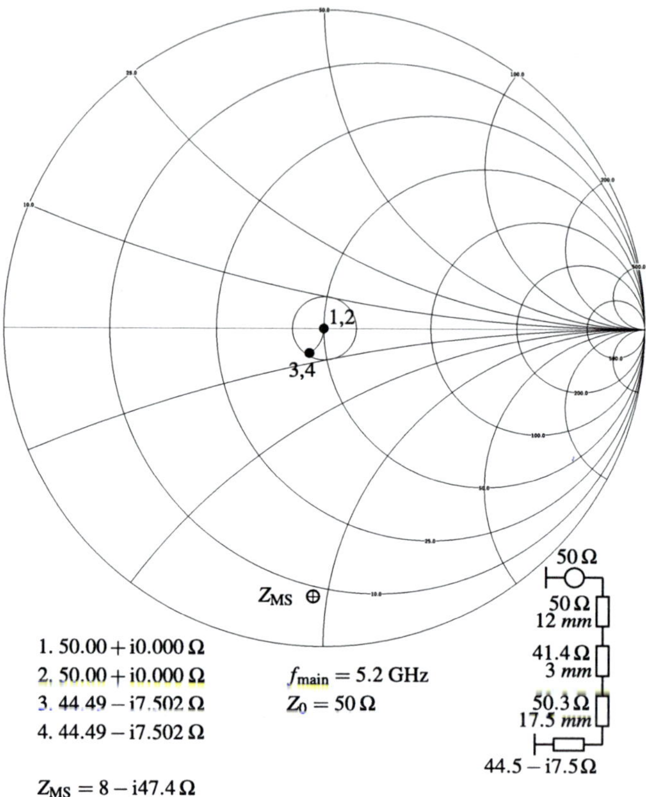

Fig. 10.12 5.2 GHz source impedance transformation caused by the embedding network. Target source impedance also shown

dered onto the board and a SOLT measurement was performed at the fixture reference planes (refer to Fig. 10.1). Figure 10.14 depicts the measured S_{11} and S_{22} parameters.

Figure 10.14 reveals that both the input and the output of the RF power amplifier are very well matched, as was expected. The small signal power gain of the matched power amplifier can be seen through its measured S_{21} characteristics depicted in Fig. 10.15. It shows that the power gain of the RF power amplifier was increased from 2.4 dB at 5.2 GHz in an unmatched condition to 8 dB at 5.2 GHz when matched using the procedure presented in this appendix. In Fig. 10.15, the reverse gain S_{12} is also shown. A summary of the measured S parameters at 5.2 GHz for the matched power amplifier is shown in Table 10.3.

10.4 Analysis of the Results

Comparing the gain achieved in the matched condition to the maximum gain calculated according to (10.7) and shown in Table 10.2, the former is only 1.1 dB

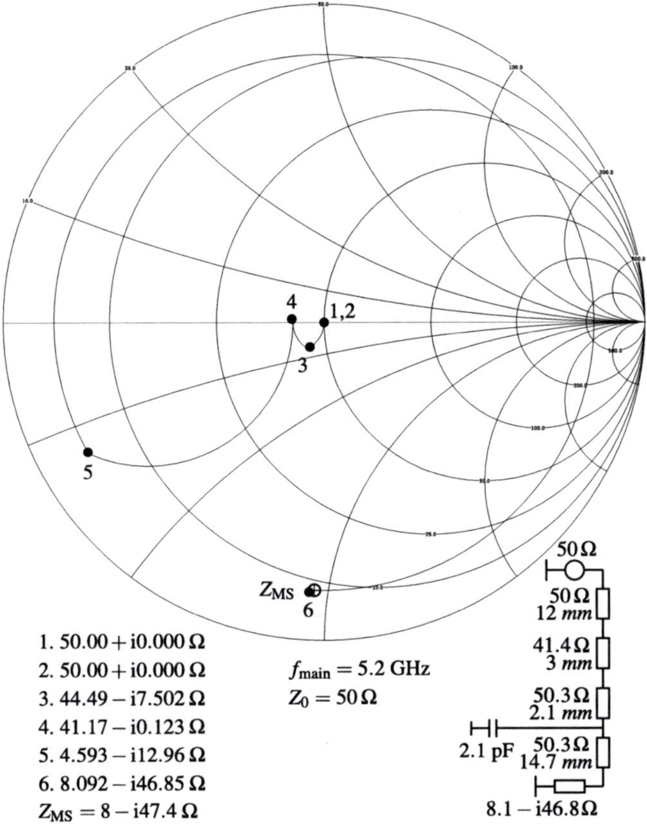

1. $50.00 + i0.000\,\Omega$
2. $50.00 + i0.000\,\Omega$
3. $44.49 - i7.502\,\Omega$
4. $41.17 - i0.123\,\Omega$
5. $4.593 - i12.96\,\Omega$
6. $8.092 - i46.85\,\Omega$
$Z_{MS} = 8 - i47.4\,\Omega$

$f_{main} = 5.2$ GHz
$Z_0 = 50\,\Omega$

Fig. 10.13 Shunt capacitor added to the input microstrip access lines between points **3** and **4** of the previous figure

Table 10.3 Summary of the measured SOLT S parameters for the matched power amplifier at 5.2 GHz

Parameter	Value	Parameter	Value
S_{11}	$0.06\angle125°$	$Z_{11}\ (\Omega)$	$54.2 + i6.3$
S_{22}	$0.07\angle53°$	$Z_{22}\ (\Omega)$	$46.4 + i4.7$
S_{11} (dB)	-22.8	S_{21} (dB)	8.04
S_{22} (dB)	-24.23	S_{12} (dB)	-16.55

below the later without taking into account the ohmic losses in the microstrip access lines. A comparison of Figs. 10.3 and 10.4 to 10.14, of Figs. 10.5 to 10.15, and of Tables 10.1 to 10.3 shows that the objective of a simultaneous conjugate match was achieved, resulting in an increase of over 5 dB in power gain. This validates the techniques described as a thorough, practical and systematic procedure that can help in the implementation of impedance matching networks for printed-circuit RF amplifiers.

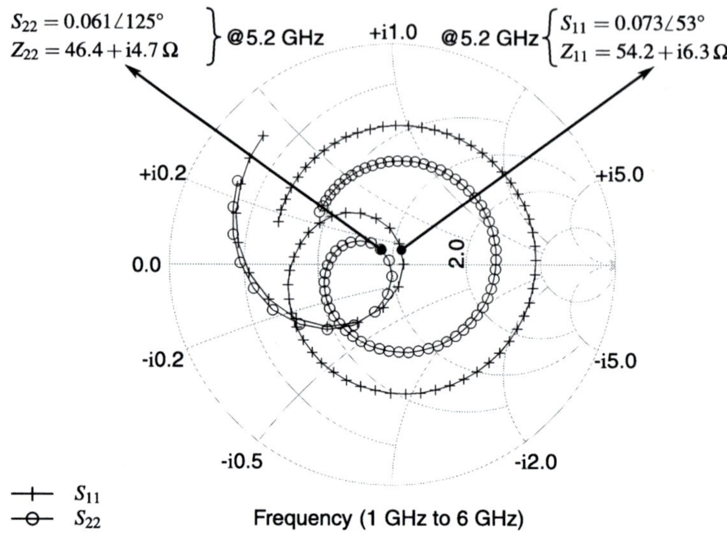

Fig. 10.14 Matched power amplifier SOLT measurement (S_{11} and S_{22})

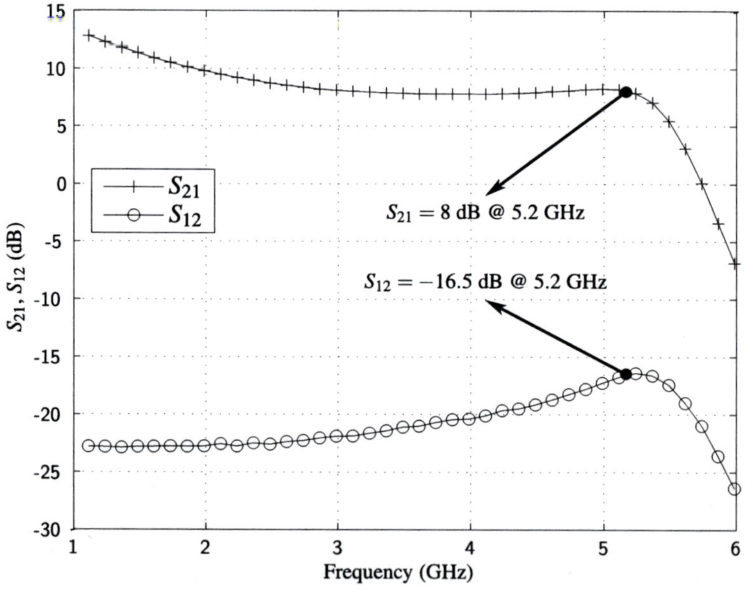

Fig. 10.15 Matched power amplifier SOLT measurement (S_{21} and S_{12})

The main advantages of the proposed impedance matching procedure, in comparison with others reported in the literature, are:

1. The need of only one step where the PCB is fabricated together with the TRL standards. The PCB is used to extract the S parameters of the DUT and the implementation of the matching networks is done in real time using the same PCB. In other approaches [9, 23, 24], two steps are required for PCB fabrication. In [24], one board is used to determine the source and load impedances and, in the second board, the matching network is implemented. In [9, 23], one PCB is used to extract the DUT S parameters and the second to implement the matching networks based on these extracted parameters. Moreover, in these three approaches, the PCBs cannot be fabricated simultaneously.
2. Optimization is not required. In the other three approaches [9, 23, 24] optimization is part of the procedure.

Furthermore, in [23], after matching an RF PA with the approach therein presented, the 5.2 dB power gain reached is 3.5 dB below the maximum theoretical gain ($G_{\max} = 8.7$ dB). With our procedure, the matched RF PA reached a power gain of 8 dB. This is only 1.1 dB below the maximum theoretical gain of 9.1 dB.

Figure 10.8 shows that using TRL and SOLT calibration techniques results in the same extracted DUT S parameters. This indicates that just one of the two calibration methods would be required. However, this is only true when the model of the embedding networks is accurate. Therefore, we recommend that both techniques are used until the accuracy of the model can be verified.

References

1. Abrie PLD (1985) The Design of Impedance-Matching Networks for Radio-Frequency and Microwave Amplifiers. Artech House, Dedham
2. ADS (2010) Advanced Design System (ADS). URL http://eesof.tm.agilent.com/products/ads_main.html
3. Agilent (1999) 8719D network analyzers. Agilent Technologies, USA. URL http://cp.literature.agilent.com/litweb/pdf/08720-90288.pdf
4. Agilent (2002) Applying error correction to network analyzer measurements. Application Note 1287-3. URL http://cp.literature.agilent.com/litweb/pdf/5965-7709E.pdf
5. Agilent (2005) Accurate measurement of packaged RF devices. White Paper. URL http://cp.literature.agilent.com/litweb/pdf/5989-3246EN.pdf
6. Agilent (2006) In-fixture measurements using vector network analyzers. Application Note 1287-9. URL http://cp.literature.agilent.com/litweb/pdf/5968-5329E.pdf
7. Agilent (2007) 85052D 3.5mm economy calibration kit. User's and Service Guide. URL http://cp.literature.agilent.com/litweb/pdf/85052-90079.pdf
8. Bauer RF, Penfield P Jr (1974) De-embedding and unterminating. IEEE Trans Microw Theory Tech MTT-22(3):282–288
9. Choi SC, Youm JE, Hwang SW (2004) Simple PCB based S-parameter extraction method for RF amplifier circuits. In: 63rd Autom RF Tech Group (ARFTG) Conf Dig, Ft Worth, TX, pp 53–59
10. Cripps SC (2006) RF Power Amplifiers for Wireless Communications, 2nd edn. Artech House, Norwood
11. Dellsperger F (2005) Smith V2.03—Software for easy circuit design with Smith Chart. URL http://fritz.dellsperger.net/
12. DuFault MD, Sharma AK (1996) A novel calibration verification procedure for millimeter-wave measurements. In: 1996 IEEE MTT-S Int Microw Symp Dig, vol 3, pp 1391–1394

13. Elmore G (1985) De-embedded measurements using the HP 8510 microwave network analyzer. In: 25th Autom RF Tech Group (ARFTG) Conf Dig, vol 7, pp 124–143
14. Engen GF, Hoer CA (1979) Thru-Reflect-Line: an improved technique for calibrating the dual six-port automatic network analyzer. IEEE Trans Microw Theory Tech MTT-27(12):987–993
15. Fitzpatrick J (1978) Error models for systems measurement. Microw J 21(5):63–66
16. Franzen NR, Speciale RA (1975) A new procedure for system calibration and error removal in automated S-parameter measurements. In: 5th European Microw Conf, pp 69–73
17. Gonzalez G (1997) Microwave Transistor Amplifiers: Analysis and Design, 2nd edn. Prentice-Hall, Upper Saddle River
18. Infineon (2007) Simple microstrip matching for all impedances. Application Note No 022. URL http://www.infineon.com
19. Lane R (1984) De-embedding device scattering parameters. Microw J 8:149–156
20. Lee TH (1998) The Design of CMOS Radio-Frequency Integrated Circuits. Cambridge University Press, Cambridge
21. Marks RB (1991) A multiline method of network analyzer calibration. IEEE Trans Microw Theory Tech 39(7):1205–1215
22. Murata (2007) Murata Chip S-parameter and Impedance Library Version 3.12.0. URL http://www.murata.com/designlib/mcsil/index.html
23. Nickel JG, Schutt-Aine JE (2003) Matched coupled microstrip transistor amplifier methodology. IEEE Trans Adv Packag 26(4):361–367
24. O'Reilly GT, Neidert RE, Wilson LK (1974) Designing microstrip matching networks for microwave-transistor power amplifiers. IEEE Trans Microw Theory Tech 22(12):1323–1325
25. Pieper R, Dellsperger F (2001) Personal computer assisted tutorial for Smith charts. In: 2001 Proc 33rd Southeast Symp Syst Theory, Athens, OH, pp 139–143
26. Pozar DM (1998) Microwave Engineering, 2nd edn. Wiley, New York
27. Rehnmark S (1974) On the calibration process of automatic network analyzer systems. IEEE Trans Microw Theory Tech 22(4):457–458
28. Rogers (2006) RO4000 series high frequency circuit materials. Data Sheet 92-004. URL http://www.rogerscorp.com/acm/literature.aspx
29. Rytting D (1998) Network analyzer error models and calibration methods. IEEE MTT/ED Seminar: Calibration and Error Correction Techniques for Network Analysis. URL http://cpd.ogi.edu/IEEE-MTT-ED/DougRyttingSeminar.htm
30. Scott JB (2005) Investigation of a method to improve VNA calibration in planar dispersive media through adding an asymmetrical reciprocal device. IEEE Trans Microw Theory Tech 53(9):3007–3013
31. Silvonen KJ (1992) A general approach to network analyzer calibration. IEEE Trans Microw Theory Tech 40(4):754–759
32. Vaitkus R, Scheitlin D (1982) A two-tier deembedding technique for packaged transistors In: 1982 IEEE MTT-S Int Microw Symp Dig, vol 82, pp 328–330
33. Vaitkus RL (1986) Wide-band de-embedding with a short, an open, and a through line. Proc IEEE 74(1):71–74
34. Wartenberg SA, Grajek P (2001) De-embedding PCB fixtures for package characterization. Microw Opt Technol Lett 31(2):111–112
35. Williams D (1990) De-embedding and unterminating microwave fixtures with nonlinear least squares. IEEE Trans Microw Theory Tech 38(6):787–791

Index

A
Active inductors, 66, 67, 72
Anti-overlapping circuit, 23

B
Balanced amplifier, 64, 65, 72
Bandwidth, 63, 78
    broadband, 63
    envelope, 9, 55
    narrowband, 63, 65
Bias adaption, 11
Bondwire, 40, 64, 110, 126, 137, 146
Breakdown voltage, 2, 26, 28
BST capacitors, 65, 66, 72

C
Calibration, 45, 102, 105, 111, 139–143, 155
Center frequency, 34, 62, 63, 102, 115
Chip-on-board, 40, 110, 138
Comparator, 18–21
Compression, 17, 18, 49, 50, 64, 72
Coupled inductors, 66, 68–71, 84–89, 96, 104, 119, 124
Coupling factor, 68, 85, 86, 101, 108, 119, 120, 126
CW, *see* single tone

D
De-embedding, 104, 108, 137–142, 147
Delay
    anti-overlapping, 23
    comparator, 20
    electrical, 142, 146, 149
    modulator, 12, 19, 20, 33, 37, 42, 125

Dielectric constant, 40, 66
Distributed amplifier, 64, 65, 72
Dynamic supply, 2–4, 10–14, 17–20, 24, 25, 28, 32–35, 37, 39–45, 49–53, 55, 56, 58, 59, 127, 131–134

E
EER, 8–10, 13, 14
Efficiency
    drain efficiency, 7
    modulator, 35, 52
    PAE, *see* PAE
Envelope detector, 10, 17, 18, 32, 41, 44, 50, 55, 58, 59, 125
Envelope following, 11
Envelope tracking, 11
EVM, 3, 6, 13, 55, 56, 58, 132, 134

F
Ferromagnetic inductors, 66–68, 72
Frequency-tunable PA, 3, 4, 61–63, 65, 66, 72, 83, 84, 89, 96, 99, 101, 108, 111, 115, 118–120, 124, 126

G
Gyrator, 67

I
IMD3, 3, 6, 11, 13, 56, 111
    dynamic supply, 32, 35–37, 49, 51, 53–55, 125, 131
    tunable, 94–99, 103, 105, 112, 115–117
Impedance matching
    conjugate match, 77, 78
    power match, 77, 78

P.A. Dal Fabbro, M. Kayal, *Linear CMOS RF Power Amplifiers for Wireless Applications,*   157
Analog Circuits and Signal Processing,
DOI 10.1007/978-90-481-9361-5, © Springer Science+Business Media B.V. 2010

Impedance matching (*cont.*)
  simultaneous conjugate, 102, 104

K
Knee voltage, 26, 28, 42, 78

L
LC filter, 12, 18, 19, 24, 25, 32–35, 37, 41, 42, 55, 58
Load pull, 94–96
Lossy matching, 63, 65, 72

M
Matching network, 26, 28, 45, 61–63, 65, 77–79, 82, 124, 138, 139, 155
  design, 78, 102, 138, 142
  designability, 93
  input, 77, 83, 95, 137, 151
  L-matching, 71, 78, 95
  output, 77, 78, 81–84, 95, 99, 124, 137, 150
  π-matching, 71, 78, 80–84, 93–99, 125
  T-matching, 78
  tunable, 66, 67, 71, 82, 83, 125, 126
MEMS inductors, 66, 67, 72
MEMS switches, 66, 71, 72
Microstrip, 41, 138–153
MOS capacitor, 65, 66, 72
Mutual inductance, 68, 86

O
OFDM, 1, 3, 50, 55, 56, 127, 132–135
Optimum resistance, 26–29, 45, 77–81, 93, 94, 96, 120

P
Package parasitics, 64, 138
PAE, 13
  definition, 7
  dynamic supply, 35–37, 52–55, 57
  tunable, 83, 96–99, 103, 105, 111, 112, 115–118
PAE ratio
  dynamic supply, 52, 53, 56
  tunable, 99, 115
PAPR, 10, 12
PCB
  dynamic supply, 39–41, 44, 50
  matching implementation, 137–155
  tunable, 110, 118

Photograph
  coupled inductors, 104, 106
  dynamic supply PA, 40, 41
  stand-alone PA, 102
  tunable PA, 108, 110, 118
PIN diodes, 66, 71, 72
Power gain, 2, 7, 17, 26, 42, 49–51, 53, 63, 65, 83, 84, 103, 145
  forward, 45, 111, 144, 145, 152, 155
  maximum, 148, 155
  reverse, 45, 111, 144
PWM, 19, 24, 32

Q
Quality factor, 64, 67, 68, 70, 71, 78, 79, 84, 87, 126
  enhancement, 70, 84, 85, 99
  loaded, 78–80, 93
  unloaded, 70, 82, 92

R
Reflection coefficient, 29, 45, 81, 111, 141, 143, 148

S
S parameters
  measurement, 45, 51, 102, 103, 105, 106, 110, 111, 127, 137, 141–145, 147–149, 152
  simulation, 30
Saturable reactor, 66, 68
SFG, 88
  DPI, 88
Single-tone, 49, 51
Slew rate, 21
Sliding-mode control, 19
SRF, 67, 86, 142
Stability, 29
  NDF, 30
  Rollett criteria, 29
  Rollett stability factor, 29
  Rollett's proviso, 30
  stability circles, 30
Switching, 66
Synchronous switch, 18, 21–23

T
Threshold voltage, 22, 66
Transformation factor, 62, 66, 78–80, 93, 94, 124
Transformer, 85, 88
   characterization, 106, 108
   design, 86
   discrete, 114, 120, 124
   Finlay, *see* stacked
   Frlan, *see* interleaved
   integrated, 85, 119, 126
   interleaved, 85, 86
   Rabjohn, 86
   stacked, 85
   stacked interleaved, 86
   T model, 106
   tapped, 85
   types, 85
Tuning range, 62, 63, 68, 71, 72, 92, 94, 114, 120
2-tone
   measurement, 3, 49–55, 59, 102, 104, 105, 111, 115, 118, 125, 127, 130, 131, 134
   simulation, 32, 34, 36, 94, 96, 99

V
Varactor, 65, 66
   varactor diode, 65–67, 72
Variable inductors, 66, 67, 72, 124
VSWR, 63, 64

W
WLAN, 1, 3, 6, 7, 12, 13, 50, 56, 58, 59, 125